D1257453

THE "APOLLO" OF AERONAUTICS

The opinions expressed in this volume are those of the authors and do not necessarily reflect the official position of the United States Government or of the National Aeronautics and Space Administration.

For sale by the Superintendent of Documents, U.S. Government Printing Office
Internet: bookstore.gpo.gov Phone: toll free (866) 512-1800; DC area (202) 512-1800
Fax: (202) 512-2104 Mail: Stop IDCC, Washington, DC 20402-0001

ISBN 978-0-16-084295-5

THE "APOLLO" OF AERONAUTICS

NASA's Aircraft Energy Efficiency Program 1973–1987

MARK D. BOWLES

National Aeronautics and Space Administration

Headquarters

300 E St SW

Washington, DC 20546

2010

SP-2009-574

www.nasa.gov

About the Cover:
Front cover: NASA Langley Research Center's Boeing 737 test aircraft on the ramp at Orlando International Airport after a day of flight tests. (NASA Langley Research Center [NASA LaRC].)
Cover design by Janine Wise.

Library of Congress Cataloging-in-Publication Data

Bowles, Mark D.
The "Apollo" of aeronautics : NASA's Aircraft Energy Efficiency Program, 1973-1987 / Mark D. Bowles.
p. cm.
Includes bibliographical references and index.
1. Aircraft Energy Efficiency Program (U.S.)--History. 2. Airplanes--Fuel consumption--Research--United States--History--20th century. 3. Aerodynamics--Research--United States--History--20th century. 4. Jet engines--Research--United States--History--20th century. I. Title.
TL704.7.B634 2009
629.134'35--dc22

2009046465

For Nancy, Isabelle, Emma, and Sarah

Table of Contents

INTRODUCTION

In fall 1975, 10 distinguished United States Senators from the Aeronautical and Space Sciences Committee summoned a group of elite aviation experts to Washington, DC. The Senators were holding hearings regarding the state of the American airline industry, which was struggling in the wake of the 1973 Arab oil embargo and the dramatically increasing cost of fuel. Providing testimony were presidents or vice presidents of United Airlines, Boeing, Pratt & Whitney, and General Electric. Other witnesses included high-ranking officials from the National Aeronautics and Space Administration (NASA), the U.S. Air Force, and the American Institute of Aeronautics and Astronautics. Their Capitol Hill testimony painted a bleak economic picture, described in phrases that included "immediate crisis condition," "long-range trouble," "serious danger," and "economic dislocation." [1] Fuel costs had recently risen from $2.59 to $11.65 for a barrel of oil and from 38.5 cents to 55.1 cents for a gallon of gasoline. While everyone knew about the increasing costs of filling up his or her own automobile, the effect on commercial aviation was taking a greater toll. The airlines industry furloughed over 25,000 employees in January 1974. Pan American, at the time the United States' largest commercial airline, suspended service to 12 cities.[2] The president of United Airlines concluded, "The economic vitality of the industry is draining away."[3]

Oil was fueling America's industrial and military might, while the majority of the world's reserves were not under United States soil. The fuel crisis of the 1970s threatened not only the airline industry but also

---

1. Statement by various participants to the Senate Committee on Aeronautical and Space Sciences, Nov. 4, 1975, Box 179, Division 8000, NASA Glenn archives.

2. "Airlines to Furlough 25,000 by January Due to Fuel Crisis; Pan Am Seeks Cutbacks," *Wall Street Journal,* Dec. 11, 1973, p. 12.

3. Statement by Charles F. McErlean to the Senate Committee on Aeronautical and Space Sciences, Nov. 4, 1975, Box 179, Division 8000, NASA Glenn archives.

the future of American prosperity itself, a situation that created a sense of panic and urgency among all Americans, from politicians on Capitol Hill to average citizens waiting in ever-longer gas lines for more expensive fuel. But the crisis also served as the genesis of technological ingenuity and innovation from a group of scientists and engineers at NASA, who initiated planning exercises to explore new fuel-saving technologies. What emerged was a series of technologically daring aeronautical programs with the potential to reduce by an astonishing 50 percent the amount of fuel used by the Nation's commercial and military aircraft. Though the endeavor was a costly 10-year, $500-million research and development (R&D) program, the United States Senators involved proclaimed that they could not "allow this technology to lie fallow."[4] The Aircraft Energy Efficiency (ACEE) project was born.

This energy crisis of the 1970s marked a turning point for the United States in a number of ways, one of which was that it changed fundamentally the focus of NASA's aeronautical research. Since its establishment in 1915 (as the National Advisory Committee for Aeronautics) and through its transformation into NASA in 1958, the organization's aeronautical emphasis had been on how to research and build aircraft that would fly higher, go faster, and travel farther.[5] "Higher, faster, and farther" were all visible aviation goals well suited for the setting of records and pushing the boundaries of engineering and piloting skill.[6] According to one aviation engineer, "The dream to fly higher, faster, and farther has driven our finest engineering and science talents to achieve what many thought was impossible."[7]

---

4. "Aircraft Fuel Efficiency Program," Report of the Committee on Aeronautical and Space Sciences of the United States Senate, Feb. 17, 1976.
5. L.W. Reithmaier, *Mach 1 and Beyond: The Illustrated Guide to High-Speed Flight* (New York: TAB Books, 1995), p. 189. Jeffrey L. Ethell, *Fuel Economy in Aviation* (Washington, DC: NASA SP-462, 1983), p. 1. Stephen L. McFarland, "Higher, Faster, and Farther: Fueling the Aeronautical Revolution, 1919–45," *Innovation and the Development of Flight*, Roger D. Launius, ed. (Texas: Texas A&M University Press, 1999), pp. 100–131.
6. After advances in speed, as well as airfoils, composite structures, and onboard computers in the 1960s and 1970s, the "era of higher, faster, and farther in flight records was largely over." Donald M. Pattilo, *Pushing the Envelope: The American Aircraft Industry* (Michigan: University of Michigan Press, 1998), p. 267.
7. Brian H. Rowe with Martin Ducheny, *The Power to Fly: An Engineer's Life* (Reston, VA: American Institute of Aeronautics and Astronautics, 2005), p. v.

The end of the first SST era (July 1, 1973). A model of the Supersonic Transport (SST) variable sweep version, with wings in the low-speed position, mounted prior to tests in the Full Scale Wind Tunnel. (NASA Langley Research Center [NASA LaRC].)

These were goals that, once achieved, could be celebrated by the public, developed by industry, and incorporated into military and commercial aviation endeavors. Sacrificing some of these capabilities in favor of fuel economy was simply unthinkable and unnecessary for roughly the first 75 years of aviation history. Fuel economy inspired no young engineers to dream impossible dreams, because fuel was simply too abundant and inexpensive to be a factor in aircraft design.

One example of what Langley engineer Joseph Chambers called the "need for speed" was the effort to create a viable supersonic civil aircraft. Business and pleasure travelers wanted to get to their destinations quickly and in comfort. The fuel efficiency of the plane they rode in rarely entered their minds. As a result, when supersonic jet technology emerged for military applications in the 1950s, managers of the commercial air transportation system dreamed of a similar model for commercial travelers: the Supersonic Transport (SST). However, these early and rushed attempts resulted in failed programs. Chambers said that was an "ill-fated national effort within the United States for an SST," which was terminated in 1971.[8]

The oil embargo in 1973 suddenly added a new focus to the aeronautical agenda and caused the United States to rethink its aviation priorities. The mantra of "higher, faster, farther" began to take a back seat to new less glamorous but more essential goals, such as conservation and efficiency. By 1976, ACEE was fully funded. Research began immediately, and it became the primary response to the Nation's crisis in the skies. ACEE consisted of six aeronautical projects divided between two NASA Centers. Three of the projects concentrated on propulsion systems, and NASA assigned its management to Lewis Research Center (now Glenn Research Center) in Cleveland, OH. These included the Engine Component Improvement project to incorporate incremental and short-term changes into existing engines to make them more efficient. The Energy Efficient Engine (E^3) project was much more daring; Lewis engineers worked toward developing an entirely new engine that promised significant fuel economies over existing turbine-powered jet engines. Most radical of all the Lewis projects was the groundbreaking Advanced Turboprop Project (ATP), an attempt at replacing the turbojet with a much more efficient propeller. Though the Advanced Turboprop did not fly as far or as fast as its jet counterpart, it could do so at vastly improved fuel efficiencies. It was Lewis's riskiest program and also most important in terms of fuel efficiency. It represented an odd confluence of old-fashioned and cutting-edge technology.

NASA assigned three other ACEE projects to Langley Research Center in Hampton, VA. The first was Energy Efficient Transport, an aerodynamics and active controls project that included a variety of initiatives to reduce drag and make flight operations more efficient. A second project was the Composite Primary Aircraft Structures, which used new materials (such as fiberglass-reinforced plastics and graphite) to replace metal and aluminum components. This significantly decreased aircraft weight

8. Joseph R. Chambers, *Innovation in Flight: Research of the NASA Langley Research Center on Revolutionary Advanced Concepts for Aeronautics* (Washington, DC: NASA History Division, 2005), p. 6. Erik Conway in has masterfully told this story. Conway, *High-Speed Dreams: NASA and the Technopolitics of Supersonic Transportation, 1945–1999* (Baltimore: Johns Hopkins University Press, 2005).

and increased fuel efficiency. A third project, Laminar Flow Control, also promised to reduce drag, and it was Langley's most challenging project of the three. NASA accepted the risk, much like the Advanced Turboprop, because Laminar Flow Control, if achieved, represented the most significant potential fuel savings of any of the ACEE programs.

NASA conducted two different types of R&D programs. The first was for "fundamental" or "base" research, where engineers conceptualized, developed, and tested initial ideas that could later lead to a successful commercial or military technology. Once these base programs reached a certain level of maturity and technological success, they were ready for the next R&D stage. This second, or "focused," R&D program typically required the allocation of large amounts of funding in order to create a full-scale demo. ACEE was an example of a "focused" program that utilized the success of existing "base" programs (such as Laminar Flow Control, winglets, and supercritical wings). The ACEE focused program and funding offered the best way to mature the fundamental technological successes already being developed.[9]

But there was a problem. The civil aircraft industry was notoriously conservative and did not often welcome change or pursue it aggressively. The ACEE program represented a dramatically different vision of future commercial flight. Although several of the programs explored slower evolutionary developments, the energy crisis inspired enough fear that the industry was willing to support the more revolutionary projects. Donald Nored, who served as director of Lewis's three ACEE projects, remarked, "The climate made people do things that normally they'd be too conservative to do."[10] The Lewis Advanced Turboprop demonstrated how a radical innovation could emerge from a dense, conservative web of bureaucracy. Its proponents thought it would revolutionize the world's aircraft propulsion systems. Likewise, Langley's programs also pushed revolutionary new technologies such as Laminar Flow Control, which many believed was impossible to achieve and foolhardy to attempt. The economics of the energy crisis shaped a climate whereby the Government, with industry encouragement and support, gave NASA the go-ahead and appropriate funding to embark upon programs that typically would have never been attempted.

---

9. Chambers, correspondence with Mark D. Bowles, Mar. 28, 2009.

10. Interview with Donald Nored, by Virginia P. Dawson and Bowles, Aug. 15, 1995.

ACEE was vitally important, and while many technical reports have been written about its programs, it has received little historical analysis. While the program appears as a footnote or sidelight in several important historical works, it is rarely placed at the forefront and given exclusive attention.[11] One exception was in 1998, when Virginia P. Dawson and Mark D. Bowles's article on the Advanced Turboprop Project in Pamela Mack's edited collection, *From Engineering Science to Big Science*.[12] The collaboration and research for that article, "Radical Innovation in a Conservative Environment," laid the groundwork for this current monograph.

Some of the best monographs and technical reports for the Lewis ACEE projects include Roy Hager and Deborah Vrabel's *Advanced Turboprop Project* and Carl C. Ciepluch's published results of the Energy Efficient Engine project.[13] Langley's ACEE projects have been the subjects of Marvin B. Dow's review of composites research, David B. Middleton's program summary of the Energy Efficient Transport project, and Albert L. Braslow's history of laminar flow control.[14] Jeffrey L. Ethell's *Fuel*

---

11. Examples include: Roger E. Bilstein, *Testing Aircraft, Exploring Space: An Illustrated History of NACA and NASA* (Baltimore: Johns Hopkins University Press, 2003), pp. 122–123; James R. Hansen, *The Bird Is on the Wing: Aerodynamics and the Progress of the American Airplane*. Centennial of Flight Series, No. 6. (College Station: Texas A&M University Press, 2004), p. 204; Louis J. Williams, *Small Transport Aircraft Technology* (The Minerva Group, 2001), p. 37; William D. Siuru and John D. Busick, *Future Flight: The Next Generation of Aircraft Technology,* (Blue Ridge Summit, PA: TAB/AERO, 1994), p. 5; Conway, *High-Speed Dreams,* p. 265; Ahmed Khairy Noor, *Structures Technology: Historical Perspective and Evolution* (Reston, VA: AIAA, 1998), p. 298.

12. Bowles and Dawson, "The Advanced Turboprop Project: Radical Innovation in a Conservative Environment," *From Engineering Science to Big Science: The NACA and NASA Collier Trophy Research Project Winners,* Pamela E. Mack, ed. (Washington, DC: NASA SP-4219, 1998), pp. 321–343.

13. Roy D. Hager and Deborah Vrabel, *Advanced Turboprop Project* (Washington, DC: NASA SP-495, 1988). Carl C. Ciepluch, Donald Y. Davis, and David E. Gray, "Results of NASA's Energy Efficient Engine Program," *Journal of Propulsion,* vol. 3, No. 6 (Nov.–Dec. 1987), pp. 560–568.

14. Marvin B. Dow, "The ACEE Program and Basic Composites Research at Langley Research Center (1975 to 1986)," (Washington, DC: NASA RP-1177, 1987). David B. Middleton, Dennis W. Bartlett, and Ray V. Hood, *Energy Efficient Transport Technology* (Washington, DC: NASA RP-1135, Sept. 1985). Albert L. Braslow, *A History of Suction-Type Laminar Flow Control with Emphasis on Flight Research* (Washington, DC: NASA *Monographs in Aerospace History* No. 13, 1999).

Economy in Aviation is an excellent technical overview of the entire ACEE program.[15] A vast number of technical reports were written during the course of these projects themselves. For example, a bibliography of the Composite Primary Aircraft Structures program alone, compiled in 1987, contains over 600 entries for technical reports just for this one ACEE program. These studies, however, focus primarily on technological evolution and achievements, and most were written while ACEE was still an active program or just shortly after its conclusion. This monograph, *The "Apollo" of Aeronautics*, examines the ACEE program more than 20 years after its termination and places it within a political, cultural, and economic context, which is absent from most of the previous work.

Taken together, the ACEE programs at Langley and Lewis represented an important moment in our technological history, which deserves further analysis for several reasons. First, it was tremendously successful on a number of technological levels. Many of the six ACEE projects led to significant improvements in fuel efficiency. One measure of this success is how much more fuel-efficient commercial airplanes are today, compared with the mid-1970s, when the ACEE program began. An estimate in 1999 suggested that aircraft energy efficiency improved on an average of 3 to 4 percent each year, and that the "world's airlines now use only about half as much fuel to carry a passenger a set distance as they did in the mid-1970s."[16] This important statistic testifies to the improved fuel efficiency stimulated by the ACEE program. While it alone was not responsible for this achievement, it served as a key industry enabler and catalyst to incorporate new fuel-savings technology into its operating fleets.

Second, ACEE represents an important case study in technology transfer to the civil industry. The goal of ACEE, from its inception, was for NASA to partner with industry to achieve a specific goal—a fuel-efficient aircraft to counteract the energy crisis. NASA, as an Agency, was important because it was able to assume the risk for technically radical projects thought to be too difficult and costly for industry alone to sponsor. "Aeronautics" was the "first A" in NASA, and this technology

15. Jeffrey L. Ethell, *Fuel Economy in Aviation* (Washington, DC: NASA SP-462, 1983).
16. Lisa Mastny, "World Air Traffic Soaring," *Vital Signs 1999: The Environmental Trends that are Shaping Our Future,* Lester Russell Brown, Michael Renner, and Linda Starke, eds. (New York: Norton, 1999), p. 86.

transfer program to the aviation industry was a way for it to reconnect with its historical roots as the National Advisory Committee for Aeronautics (NACA). This also offered a way for NASA to prove it still could make vital contributions to aeronautics research and, at the same time, demonstrated the successful "focused" R&D approach to maturing technology versus the attempt to advance technology with low-level "base" programs. The ACEE program also exemplified one way in which NASA turned its sights Earthward after the golden age of Moon landings and focused on energy conservation—an issue that continues to be of increasing importance in the 21st century.

Third, the history of this program represents an important case study in technological creativity and risk, a theme highlighted in the Dawson and Bowles article on the Advanced Turboprop Project. Thomas Hughes, a prominent historian of technology, has argued that the research and development organizations of the 20th century, regardless of whether they are run by a Government, industry, or members of a university community, stifle technical creativity. "Organizations did not support the radical inventions of the detached independent inventor," Hughes wrote, "because, like radical ideas in general, they upset the old, or introduced a new status quo."[17] In contrast, the late 19th century for Hughes was the "golden era" of invention—a time when the independent inventor flourished without institutional constraints. Historian David Hounshell has challenged Hughes's contention that industrial research laboratories "exploit creative, inventive geniuses; they neither produce nor nurture them."[18] Not only can the industrial research laboratory nurture a creative individual, but collectively, people engaged in research and development can inspire revolutionary new technological opportunities.[19]

The ACEE program represents a case in which organizational capabilities, not individual genius alone, created an opportunity for significant innovation. The organizational structure of the ACEE program (focused R&D funding) encompassed not just the various NASA Centers but also

17. Thomas Hughes, *American Genesis,* p. 54.

18. David A. Hounshell, "Hughesian History of Technology and Chandlerian Business History: Parallels, Departures, and Critics," *History and Technology,* vol. 12 (1995), p. 217.

19. Ibid. See also Hounshell and John Kenly Smith, Jr., *Science and Corporate Strategy: Du Pont R&D, 1902–1980* (Cambridge: Cambridge University Press, 1988).

included a web of industrial contracts that made it far more complex than the research laboratory of a single industrial firm. Yet the bureaucratically complex ACEE program responded to the energy crisis in an efficient way to advance some very revolutionary ideas. However, although NASA provided the environment and support for radically innovative technologies, in the end, the more conservative the ACEE technology, the more likely it was to become a commercial reality. The Lewis Advanced Turboprop Project and the Langley Laminar Flow Control program each represented the most significant fuel savings and were considered the most revolutionary of all the technologies explored. Although both were demonstrated to be technically feasible under the ACEE program, neither achieved commercial success. They were the programs most susceptible to industry neglect when the energy crisis of the 1970s subsided and fuel prices decreased.

Fourth, ACEE represents an interesting historical moment that marked a transition point between American domination of the world's civil aviation industry and rising challenges from foreign competitors, such as Airbus. From its inception, the aviation industry was different from the auto industry because the U.S. Government provided it with massive support in the name of national defense. For example, during World War II, the Government purchased planes from private manufacturers. After the war, it committed billions to finance the development of new aeronautics technology for both military and commercial aircraft. As a result of these efforts, the United States captured 90 percent of the world market and held this commanding position through the 1970s. Boeing, as the largest exporter, played a significant role in the American economy. Robert Leonard, Langley's ACEE project manager said that each jumbo jet manufactured in the United States and sold abroad offset the importation of roughly 9,000 automobiles.[20] This trade balance was vital for the United States to maintain—but it did not.

Challengers such as Airbus waited in the wings, backed with governmental commitments that far exceeded American support to the industry. Founded in 1965 as a consortium of European countries, Airbus used massive government subsidies and private investors to develop its first plane.

---

20. Robert W. Leonard, "Fuel Efficiency Through New Airframe Technology," NASA TM-84548, Aug. 1982.

The wings were made in Britain, the cockpit in France, the tail in Spain, the fuselage in West Germany, and the edge flaps in Belgium. The engines came from America. The Airbus had important innovations: it flew on two engines with two pilots, instead of three engines and three pilots. This reduced fuel consumption and lowered per flight operating costs. In 1988, Airbus captured 23 percent of the world market, and in 1999, for the first time, it received more orders for airplanes than Boeing did.[21] The story of ACEE fits within this global context and challenge, and it demonstrates the significance of the decisions made in the development and support of the next-generation aeronautical innovations. Some have argued that the same conservatism and risk aversion that defined the American civil aviation industry and enabled a challenger to take over the leading position in world market share now threatens Airbus. Today, Airbus and Boeing both face new sources of competition in Japan and China.[22]

Finally, ACEE is important because it took center stage in NASA's civil aeronautical research agenda. This led some to argue that ACEE was "the most important program in aeronautical technology in NASA" in the 1970s and 1980s.[23] Others called it the "best program NASA has had in the last ten years from the aeronautics standpoint."[24] Individual awards also attested to its importance. As a measure of the high regard of the aeronautical community, the Advanced Turboprop eventually earned a Collier Trophy, considered the most prestigious award for aerospace achievement in the United States. NASA's heritage and tradition were in aeronautics, and for ACEE to be considered the most important of all the programs put it into elite company.

Raymond Colladay, a former president of Lockheed Martin, Director of the Defense Advanced Research Projects Agency (DARPA), and NASA Associate Administrator said, "By most any metric you would use, I'd have to say yes, that it was the most important." It was important because

---

21. Robert K. Schaeffer, *Understanding Globalization: The Social Consequences of Political, Economic, and Environmental Change* (Lanham, MD: Rowman & Littlefield Publishers, 2009), pp. 9–10.
22. John Newhouse, *Boeing Versus Airbus: The Inside Story of the Greatest International Competition in Business* (New York: A.A. Knopf, 2007).
23.James J. Kramer, "Aeronautical Component Research," *Advisory Group for Aerospace Research and Development* (Report No. 782, 1990).
24."Survey Finds Little Impact of Election on Aerospace," *Aviation Week & Space Technology,* Nov. 3, 1980, p. 34.

it brought together a broad range of research and technology development programs that directly addressed a national need. According to Colladay:

"It had enough resources to make a difference, to really move things forward, whereas most of the time in the NASA aeronautics program, there isn't enough critical mass and resolve and focus to really make significant results in a timely fashion. The ACEE program did that."[25] The accompanying list highlights NASA's aeronautical programs as of 1983, which was the midpoint of the ACEE program and the 25th anniversary of NASA. While these 12 main aeronautics programs were important, ACEE contributed another 6 separate programs on its own and was generally considered the most vital for NASA in terms of technological potential and national need.) According to Joseph Chambers, it represented a perfect mixture of "funding, world economics, and technology readiness," in contrast to other programs, like the Supersonic Transport, that "spent much more money without significant impact on commercial aviation."[26]

A snapshot of NASA's aeronautical programs in 1983 is as follows:

- Supersonic Cruise Aircraft Research (SCAR) developed supersonic technologies.
- for increased range, more passengers, lighter weight, and more efficient engines.
- The Terminal-Configured Vehicle (TCV) studied problems such as landing aircraft in inclement weather, high-density traffic, aircraft noise, and takeoff and landing in highly populated areas.
- Lifting bodies were experimental wingless aircraft.
- Oblique wings were aircraft wings that could pivot 60 degrees to improve fuel efficiency.
- The Highly Maneuverable Aircraft Technology (HiMAT) was a joint program with the Air Force to test advanced fighter aircraft technologies.
- The forward-swept wing (FSW) was a joint program with DARPA to test an unusual configuration with the

25. Raymond Colladay, interview with Bowles, July 21, 2008.
26. Other examples besides the Supersonic Transport were SCAR and HSR. Chambers, correspondence with Bowles, Mar. 28, 2009.

wings swept forward 30 degrees from the fuselage for greater transonic maneuverability and better low-speed performance.

- The Quiet, Clean Short Haul Experimental Engine (QCSEE) promised lower noise and reduced emissions.
- The Quiet, Clean, General Aviation Turbofan Engine (QCGAT) looked at ways to reduce noise and emission levels for business jets.
- The Quiet Short-haul Research Aircraft (QSRA) was an experimental vehicle to investigate commercial short take-off and landing to assist in reducing airport congestion.
- The Vertical/Short Take-Off and Landing research program (V/STOL) was one of NASA's helicopter projects.
- The Rotor Systems Research Aircraft (RSRA) was another experimental helicopter that had wings.
- The Tilt Rotor Research Aircraft (TRRA) flew twice as fast as conventional helicopters and had potential military and commercial opportunities.

It became the "Apollo of Aeronautics," which represents, on the one hand, its importance, with the comparison to the visible technological icon. On the other hand, it demonstrates the longstanding belief that NASA's aeronautics mission had become a handmaiden to its space activities, existing in Apollo's shadow. The terminology refers to President Richard M. Nixon's 1973 speech in which he established a "Project Independence," with a goal of attaining energy independence for the United States by 1980. Abe Silverstein, a former Lewis Director, had written a letter to NASA Administrator George Low, discussing the President's "need for 'Apollo' type programs" for energy, efficiency, and conservation projects.[27] Silverstein had a unique connection with the name, as he was the one who suggested that NASA's missions to the Moon be called "Apollo."[28.]

---

27. Abe Silverstein to George Low, Nov. 16, 1973, Box 181, Division 8000, NASA Glenn archives.
28. Wolfgang Saxon, "Abe Silverstein, 92, Engineer Who Named Apollo Program," *New York Times,* June 5, 2001.

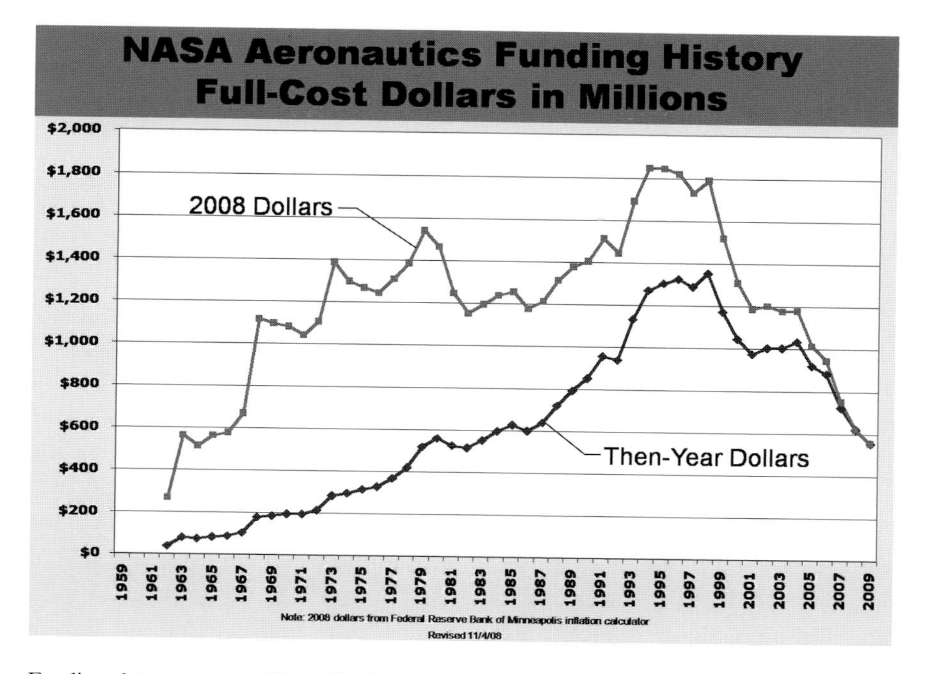

Funding data courtesy of Roy Harris, technical adviser to the NASA aeronautics support team.

The name-association between this space program and aeronautics continued with the Advanced Transport Technology program, an ACEE predecessor of the early 1970s, which Jeffrey Ethell called the "Apollo program of aeronautics."[29] Just like ACEE, it incorporated several advanced aeronautical concepts under one initiative.

But, if the best way to describe the Nation's most significant aeronautics project was through comparison to an aerospace program and the glory years of Apollo, then this cornerstone of NASA was in trouble. This was both perception and reality. Not only was ACEE threatened with cancellation just a few years after it began, but in the early 1980s, NASA had to fight to ensure that it would be allowed to keep all of its aeronautics programs under its research umbrella. Key advisers in the new Reagan administration called for the end of aeronautics for NASA. To keep it alive, NASA had to become active advocates and sellers of its aeronautical expertise to convince the Government and the Office of Management and

29. Ethell, *Fuel Economy in Aviation* (Washington, DC: NASA SP-462, 1983), p. 2.

Budget that it was in the best interest of the Nation to continue these activities. Although a stay of execution was granted, the aeronautics advocates were not entirely successful. In 1980, NASA's entire budget was $3.6 billion, with aeronautics representing just $300 million.[30] These problems for the aeronautics program continue today.

While ACEE could be considered the Apollo of fuel-conservation projects, it was also fundamentally different. Although ACEE and Apollo both responded to a national need, ACEE was unlike Apollo in that the space program had a significant development component coupled with its mission of building a spacecraft to send men to the Moon. Apollo responded to a national security and military threat. ACEE's mission was never to build an aircraft but to establish enabling technology that airline manufacturers could commercialize at their own expense. ACEE responded to an economic threat. Apollo was a single program, funded in the billions of dollars. ACEE was a series of six programs, which combined received less than a half billion dollars of funding. Despite these differences, Apollo was the pinnacle of NASA's aerospace program and became not only a symbol of the Agency's ability and excellence but also of American technical ingenuity and ability. Although the analogy is not exactly correct, ACEE was NASA's most important aeronautics program, and so it became the "Apollo of Aeronautics," fighting to emerge from Apollo's shadow.

Apollo has become *the* symbol of American achievement, as demonstrated through that familiar phrase that begins, "If we can put a man on the Moon. . . . " Though it is equally impressive that humans have been flying in the sky for just over a century, this feat no longer is as wondrous as it once was. In an era when aeronautics research is continually threatened by funding cuts and disinterest, this monograph's intentionally ironic title serves as a reminder that aeronautics research it needs to overcome its secondary status and reclaim some of its former prestige. The story told here will demonstrate the significance aeronautics has played and continues to play in the history of the United States.

Finally, this story of the ACEE program takes on special resonance when we reflect from a 21st century perspective, as hybrid technology and fuel efficiency once again become cherished commodities. In

30. General Accounting Office, "Preliminary Draft of a Proposed Report, Review of NASA's Aircraft Energy Efficiency Project," Box 182, Division 8000, NASA Glenn archives.

summer 2008, when gasoline prices were measured by increasing dollars, as opposed to increasing cents as in the early 1970s, the United States began awaking from the collective amnesia over fuel dependence suffered throughout the 1990s and the age of the SUV. In the absence of any current coordinated national effort uniting Government, industry, and academia, the successes and lessons of the ACEE program become ever more important. This $500-million program, funded by the Government, has achieved many of its goals, making the aircraft flying today significantly more fuel efficient. But the structure of the ACEE program, coupled with the willingness of the United States Government to invest in researching risky technological ideas, is what serves as a lesson today. It also serves as a warning for the consequences of failing to utilize aggressively the most revolutionary fuel-efficient technology. Government and industry left some of the most advanced ACEE fuel conservation concepts on the design table and the test stand, never integrating them into commercial flight because of decreasing oil prices (a temporary phenomenon). Today, NASA is scrambling to resurrect some of these concepts—it is, for example, now attempting to breathe new life into the Advanced Turboprop, a program long thought dead. While ACEE can be examined as a model for how to respond to the energy crisis, which continues to threaten American prosperity, it also demonstrates the consequences of technological innovation left to subsequent neglect.

ACKNOWLEDGMENTS

I first would like to thank Virginia Dawson for serving as a mentor for so many years. She first invited me to work with her as a coauthor on an article on the Advanced Turboprop Project in 1995.[31] This was the start of a long collaboration with her that blossomed into many historical projects at History Enterprises, Inc. Dr. Dawson and her expert historical work on Lewis Research Center have been and continue to be an inspiration. She also provided her insight on a draft of this monograph. I would like to thank Joseph R. Chambers for providing points on contact on the ACEE program at the Langley Research Center and for his careful reading of this manuscript. The monograph is vastly improved because of the assistance of Dr. Dawson and Mr. Chambers.

31. Bowles and Dawson, "The Advanced Turboprop Project," pp. 321–343.

Gail S. Langevin helped to coordinate my research trip to Langley and ensure I had everything I needed during my stay. The Langley library staff was supportive and assisted in the location of key materials. The archivists and librarians at Glenn Research Center were of tremendous assistance with my numerous requests. I would like to sincerely thank Kevin Coleman, Robert Arrighi, Deborah Demoline, Anne Powers, Suzanne Kelley, and Jan Dick for all their assistance over the years.

I am grateful to Elizabeth Armstrong for her expert assistance in shaping the manuscript. I also want to thank the Communications Support Services Center team at NASA Headquarters for its contributions, including Greg Finley for copyediting, Stacie Dapoz for proofreading, and Janine Wise for design. Heidi Blough expertly compiled the index.

I also appreciate those who offered their time to be interviewed for this project, including Herm Rediess, Richard Wagner, Raymond Colladay, John Klineberg, and Dennis Huff. Dr. Dawson and I also conducted interviews with Donald Nored, G. Keith Sievers, and Daniel Mikkelson for a previous article we published on the Advanced Turboprop Project. The efforts of Donald Nored in preserving the rich documentation associated with the ACEE project were essential in helping me to reconstruct this story. Without his archival spirit and sense of historical importance, many of the details would have been lost. Finally, I want to thank those who commissioned this series of aerospace monographs. They include Anthony Springer from NASA, and Lynn Bondurant, Gail Doleman Smith, and Dorothy Watkins, from Paragon Tec. NASA's willingness to engage historians and the academic freedom it ensures them is always a pleasure

I also wanted to thank Dr. Jon Carleton, the department chair of history and military studies at American Public University System. It has been a highlight of my academic life to be a member of his faculty and to engage with students from around the world on a daily basis.

No acknowledgment would be complete without recognizing the love of my family, which is the single most important force in my life.

My wife of 19 years, Nancy, has been a cherished partner of mine in anything worthwhile that I might have done in life. We have a 9-year-old daughter, Isabelle, who each day finds new ways to enlighten and enrich our lives. In February 2009, we welcomed twin girls to this world—Emma and Sarah. It is to the four of them that this book is dedicated.

Chapter 1:

Oil as a Weapon

On October 6, 1973, a terrorist's bomb shattered the solemn spiritual calm of Yom Kippur, the most sacred of holy days on the Hebrew calendar. A grenade, thrown by someone whom American newspapers referred to as an "Arab guerilla," wounded a soldier and two policemen in Israeli-occupied Gaza City.[1] This was the opening salvo of a massive, coordinated surprise attack on Israel by Egypt and Syria, whose forces crossed the Suez Canal in retaliation for the loss of their land in the Sinai and the Golan Heights during 1967's Six Day War. Israel quickly mobilized for war. Prime Minister Golda Meir proclaimed the attack an "act of madness." Her Defense Minister, Moshe Dayan, spoke in starker terms, calling for "all out war" and with a promise that "We will annihilate them."[2]

Meanwhile, Israel observed this holiest of days as a nation; its citizens spent the day fasting and praying, not listening to the radio or reading newspapers. Many had no idea the attack had occurred until they gathered for Yom Kippur services later that evening. At synagogues throughout the country, rabbis read aloud the names of those being summoned immediately to fight. In one crowded synagogue, a reservist soldier stood as his name was read, and as he turned to leave, his weeping father held him in a tight embrace, refusing to let him go. The rabbi intervened, saying, "His place is not here today." The rabbi blessed the soldier as his father released him. The Yom Kippur War (or as some called it the October War) had begun.

Over the next 3 weeks, the world witnessed combat whose intensity rivaled that of World War II. With Americans helping to arm Israel and the Soviet Union stockpiling weapons in the Arab nations, some speculated that the next world war was imminent. This did not happen, but the events

---

1."Terrorist's Bomb Opens Yom Kippur," *Los Angeles Times*, Oct. 6, 1973, p. 13.

2."Sirens Break Solemnity of Israel's Yom Kippur," *Los Angeles Times*, Oct. 7, 1973, p. 8.

of that day affected the lives of all Americans, because of a devastating economic—not military—weapon. The Arab nations retaliated against the West with an oil embargo, dramatically raising the price of oil and reducing the supply. It revealed a significant weakness of the United States, one that demonstrated how closely its economy was aligned with the accessibility of oil. Many believe this 1973 confrontation to be the genesis of the 1970s energy crisis. Although it played a major role, the crisis was actually rooted in earlier events.

The *New York Times* first used of the term "energy crisis" in relationship to the United States in 1971. In a three-part series titled "Nation's Energy Crisis," reporter John Noble Wilford recounted the effects of a Faustian bargain reaching back to the dawn of the Industrial Revolution. Dr. Faust, of German legend, was an astrologer and alchemist who sought forbidden knowledge and ultimately sold his soul to the devil, Mephistopheles, to attain it. The story has been used as a symbol for Western civilization's constant pursuit of power and knowledge.[3] Wilford used it to describe America's situation in 1971. Symbolically, the 19 century's bucolic environment was sacrificed for "modern man to command . . . and to harness in the Saturn 5 moon rocket the power of 900,000 horses."[4] This energy-dependent society had struck the Faustian energy bargain, and, Wilford argued, it resulted in the energy crisis of the 1970s.

Aside from the environmental damage wrought by industrial society, there was also the problem of how to sustain its momentum. The power to drive modern American society is derived in large part from natural resources not within its control. At the time of Wilford's article, petroleum represented 43 percent of all domestic energy usage. With more than 90 percent of all the oil consumed in the eastern half of the United States coming from sources abroad, Wilford argued, "This gives a number of foreign governments a major voice in the price and flow of American fuel."[5] And more than prices were under their control. As one geologist wrote in 1976,

---

3. Glenn Blackburn, *Western Civilization: From Early Societies to the Present* (New York: St. Martin's Press, 1991), p. 238.
4. John Noble Wilford, "Nation's Energy Crisis: It Won't Go Away Soon," *New York Times*, July 6, 1971, p. 1.
5. Wilford, "Nation's Energy Crisis: Nuclear Future Looms," *New York Times*, July 7, 1971, p. 1.

"Whoever controls the energy systems can dominate the society."[6] As the United States became a superpower in the 20th century, the American engine became increasingly powered by a fuel not of its own making. The effects of external control became evident with the onset of the Arab oil embargo.

In 1973, 2 years after the suggestion that the United States was suffering from or had an energy crisis, the Arab world began using "oil as a weapon."[7] The statistical results of the Yom Kippur War included the loss of more than 3,000 lives, as well as billions of dollars expended in military equipment. But for the first time, a new weapon emerged that had the power to destabilize all industrial nations—oil. Because of American support of Israel, Saudi Arabia announced a 10-percent reduction in the flow of oil to the United States and its allies, with the threat of an additional 5-percent reduction each month unless the West stopped sending arms to Israel. Saudi Arabia was at the time producing 8½ million barrels of oil a day, and it represented the third largest oil exporter to America, roughly 400,000 barrels per day.[8] Similar threats came from other members of the oil cartel, known as the Organization of the Petroleum Exporting Countries (OPEC). Oil was the lifeblood of the United States, and a shortage or a threat to its access quickly revealed it to be the Nation's Achilles' heel.

Although a cease-fire was negotiated by October 22, just weeks after fighting began, the conflict caused an economic ripple effect that spread throughout the world. Neither Israel nor the Arab nations officially won, but the conflict became an important symbol of national identity and strength in the Muslim world. It marked the first time Egyptian soldiers inflicted losses against Israel and won substantial territorial gains. "Crossing the Suez Canal" became a slogan that contributed to a new Arab unity and pride.[9] The conflict was significant outside the Middle East as well. Superpower patrons had come to the support of both sides, threatening to engulf the world in conflict, and it all but destroyed the détente negotiated by President Richard M. Nixon and Leonid Brezhnev. For the first time, both superpowers had a direct confrontation in the Middle East, and it served to heighten the Cold War's intensity, renewing

---

6. Earl Cook, *Man, Energy, Society* (San Francisco, CA: W.H. Freeman, 1976), p. 208.
7. Clyde H. Farnsworth, "Oil as an Arab Weapon," *New York Times*, Oct. 18, 1973, p. 97.
8. "Saudis Threaten U.S. Oil Embargo," *Washington Post*, Oct. 19, 1973, p. A1.
9. Don Peretz, *The Middle East Today*, 5th ed. (New York: Praeger, 1988), p. 254.

the United States' perceived urgency to match and surpass all Soviet military capabilities.[10]

While the United States was confident it could maintain its pace in the arms race, its leaders recognized a more significant threat in its vulnerability to the "oil weapon." In response, just 1 month after the Yom Kippur War, President Nixon signed the Alaska pipeline bill, allocating $4.5 billion to open the most significant oil reserves in the United States.[11] Soon, the Nation's speed limit would be reduced to 55 miles per hour. But oil from Alaska and slower driving would not solve the immediate crisis. A reporter from the *Washington Post* called the oil embargo the "biggest, most painful single problem ever met by the U.S. in peacetime." And many speculated that the stakes could not be higher: "The political and strategic independence of the United States" was being threatened by "oil blackmail."[12]

By March 1974, OPEC had decreased oil exports by 15 percent (a reduction of 1.5 million barrels a day to the United States) and dramatically increased prices. On March 17, the embargo essentially ended, but the impact continued to be felt. One reporter noted that the "oil weapon (which looks more like a shotgun than a rifle) has hit its target." America suffered worsening inflation, decreasing growth, and continued high oil prices. Many predicted that the "United States economy—and indeed the world economy—will never again be the same as in the pre-embargo days."[13]

The embargo itself was not responsible for the energy crisis, and its end did not make the country less vulnerable. Thomas Rees, a Democratic Congressman from California, stated that the end of the embargo would mean little security for the United States. "Having the right to buy Arab oil is having the right to go bankrupt." He continued his warning, saying, "It's not the lack of oil that will ruin the world—it's the price of oil."[14]

---

10. Victor Israelyan, *Inside the Kremlin During the Yom Kippur War* (University Park, PA: Pennsylvania State University Press, 1995), p. x.

11. "Nixon Sees 'Possibility' of Arabs Lifting Oil Embargo; Signs Alaska Pipeline Bill," *Los Angeles Times*, Nov. 16, 1973.

12. Joseph Alsop, "Oil Blackmail Threatens U.S. Independence," *Washington Post*, Nov. 21, 1973, p. A19.

13. Soma Golden, "Impact of Embargo Lingers On," *New York Times*, Mar. 17, 1974, p. 155.

14. Thomas Rees, quoted in Martha L. Willman, "Oil Embargo End May Be Ruinous," *Los Angeles Times*, Feb. 10, 1974, p. SF_C2.

The price of a barrel of oil in the past year alone had increased 450 percent, from $2.59 to $11.65. The price of a gallon of gasoline increased from 38.5 cents in May 1973 to 55.1 cents in June 1974. There seemed to be only one immediate answer to the problem—conservation.

Many oil industry experts thought that the United States could "get by without the Arab oil imports primarily by reducing American consumption."[15] As early as 1973, vocal proponents called for "strong conversation measures," including new technologies, which would reduce America's energy dependence and lessen the effectiveness of the Arabnations' oil weapon in "political and economic warfare."[16] A conservation strategy became the primary means to counter the effects of the energy crisis. Despite the urgency, it would be nearly 2 years after the oil embargo before American politicians began to pursue actively a solution to the problem.

To bring attention to this negligence, on January 29, 1975, a group of American scientists that included 11 Nobel Prize winners published a dire warning: the U.S. was facing "the most serious situation since World War II." The threat was the "energy crisis," and the group believed that the country was "courting energy disaster" through its lethargy, ignorance, and confusion.[17] Nobel laureate Hans A. Bethe, a Cornell University physicist, drafted the report. He and his colleagues warned that "our whole mode of life may come to an end unless we find a solution."[18] In agreement with Bethe's analysis, some reporters chastised the U.S. Government for "fiddling while the energy runs out."[19]

While President Gerald R. Ford had recently devised an energy program that included a $1-per-barrel excise tax on foreign oil, most believed his modest research initiatives would be ineffective. Even Ford was critical of the U.S. Congress and its lack of action on this increasingly important issue. He thought his excise tax proposal would in a sense, be "putting a gun to Congress's head," to try to motivate it to propose a plan to solve

---

15. "Saudis Threaten U.S. Oil Embargo," *Washington Post*, Oct. 19, 1973, p. A1.
16. "The Arab Oil Threat," *New York Times*, Nov. 23, 1973, p. 34.
17. Robert C. Cowen, "Fiddling While the Energy Runs Out," *Christian Science Monitor*, Jan. 29, 1975, p. 11.
18. Victor K. McElheny, "Hans Bethe Urges U.S. Drive for Atom Power and Coal," *New York Times*, Dec. 14, 1974, p. 58.
19. Cowen, "Fiddling While the Energy Runs Out," p. 11.

President Gerald R. Ford meets with Soviet and American space leaders to examine the Soviet Soyuz spacecraft model from a model set depicting the 1975 Apollo Soyuz Test Project, an Earth orbital docking and rendezvous mission with crewmen from the U.S. and U.S.S.R. (September 7, 1974). (NASA Johnson Space Center [NASA JSC].)

the crisis. On January 31, 1975, a *New York Times* reporter wrote that "the country may well be hastened into action."[20] That same day, most likely unknown to the press, to Bethe, or even to President Ford, two prominent U.S. Senators—Frank E. Moss, a Democratic Senator from Utah, and Barry Goldwater, a Republican Senator from Arizona—sent a letter to the NASA's Administrator, James C. Fletcher. They thought the Government Agency that had most recently held the Nation's attention with its successes on the surface of the Moon might have the technological capability to coordinate a major conservation initiative on Earth. This letter, dated January 31, 1975, was the genesis of what became one of the largest coordinated environmental programs ever attempted in the United States. NASA refocused its sights from the heavens to Earth.

20. Tom Wicker, "Mr. Ford Acts Like a President," *New York Times*, Jan. 31, 1975.

From the Moon to Earth

In December 1972, the last two astronauts to walk the surface of the Moon left their desolate surroundings and returned to Earth. Apollo 17 brought to an end a dramatic era at NASA that began with Kennedy's famous proclamation promising to send a man to the Moon. During the Apollo years, NASA enjoyed the world's praise as the pinnacle of humanity's technological excellence. But Apollo 17's return marked a new era. Its return signified the beginnings of a fundamental transformation in the Agency's vision, away from space and lunar exploration and toward Earth and low-Earth orbit. Astronauts would venture no further than the low-Earth destinations of the Space Shuttle, and more pressing national concerns took the focus and initiative away from long-term dreams in space. Furthermore, space initiatives had become the prime focus of NASA during the Apollo era, to the detriment of its work in aeronautics. Basic research in aeronautics was an area that many believed had been neglected for too long. A 1976 Senate Committee on Aeronautical and Space Sciences report stated "We are concerned that the nation's aeronautical research and technology base in aeronautics has in fact eroded significantly over the last several years."[21]

One of the practical earthly problems that entered NASA's new aeronautical consciousness was the energy crisis. The crisis threatened to shake the foundations of commercial flight. Prior to 1972, fuel represented one-quarter of the operating costs of a typical airline organization.[22] After 1972, foreign petroleum dependency increased, and fuel doubled its revenue drain, resulting in the reduction of flights, grounding of aircraft, and layoffs of thousands employees.[23] The situation appeared to grow worse by the day.

Because of NASA's expertise in aeronautics, the United States Congress looked to it to lead a new conservation initiative. It began with the letter that Senators Moss and Goldwater wrote January 31, 1975, to James C. Fletcher, the NASA Administrator. Although the letter came from the Senators, its origins were actually in NASA

---

21. Eugene Kozicharow, "New NASA Aeronautics Stress Sought," *Aviation Week & Space Technology*, Oct. 4, 1976, p. 23.

22. "Aircraft Fuel Conservation Technology Task Force Report," Office of Aeronautics and Space Technology, Sept. 10, 1975, p. 7.

23. Statement by Clifton F. von Kann to the Senate Committee on Aeronautical and Space Sciences, Sept. 10, 1975, Box 179, Division 8000, NASA Glenn archives.

John Klineberg spent 25 years working at NASA. He was the Director of both the Goddard Space Flight Center and Lewis (now Glenn) Research Center. He served as Deputy Associate Administrator for Aeronautics and Space Technology at Headquarters and was a research scientist at the Ames Research Center. (NASA Glenn Research Center [NASA GRC].)

itself. John Klineberg, who served NASA in a variety of leadership positions, such as head of Lewis Research Center, recently recalled "Moss, of course, wrote us a letter that justified it . . . [but] I wrote that letter."[24]

---

24. John Klineberg, interview with Bowles, July 28, 2008.

Richard T. Whitcomb examines a model incorporating his famous transonic area rule in the 8-foot High-Speed Tunnel in April 1954. (NASA Langley Research Center [NASA LaRC].)

In the letter, Moss and Goldwater said it was their desire, as leaders of the Committee on Aeronautical and Space Sciences, that NASA devise a plan to develop new technologies to lessen the effects of the energy crisis.

The plan was needed for the "preservation of the role of the United States as a leader in aeronautical science and technology." They envisioned a program led by NASA that would result in significant technology transfers to industry. NASA was to research a new generation of fuel-efficient aircraft that would cost roughly the same as current aircraft, have the same performance capabilities, meet the same safety and environmental requirements, offer significant fuel savings, and be able to take to the skies in the 1980s. Moss and Goldwater ended their letter by stating, "It is our hope that the goal you establish will be one that is both feasible and challenging."[25] Risk and the acceptance of challenge were approved and encouraged components of the daring project from the start.

NASA responded quickly to the request, in part because the Agency had already been investigating some fuel-efficient technologies as part of its base R&D activities. One of the first was the "supercritical wing," a project led by Langley's Richard Whitcomb in the mid-1960s, which delayed the formation of a shock wave until the aircraft attained a faster speed.[26]

The result was a significant cruise performance improvement and an increase in fuel efficiency. In mid-1970, NASA established the Advanced Transport Technology office to take advantage of the aerodynamic potential of the supercritical wing for flight efficiency. Other fuel-efficient technology programs were soon added. This included an Active Controls program, which used computers to control airplane surfaces to reduce drag and increase efficiency. Composite materials were also studied because of the light weight and strength of polymers compared with existing aluminum and metal airplane components.

With the oil embargo in 1973 and the resulting energy crisis, NASA intensified its explorations into this area. It established the Energy Trends and Alternative Fuels (ETAF) program in April 1973 to search for more efficient uses of petroleum and also for alternative energy sources such as hydrogen and electric power. By the end of the year, a NASA manager wrote, "The relevance and urgency of this study has grown dramatically since spring."[27] In 1973, NASA also collaborated with Hamilton Standard in a program called Reducing the Energy Consumption of Commercial Air Transportation (RECAT). Over the next 2 years, NASA, in collaboration with General Electric, Pratt & Whitney, Hamilton Standard, and

25. Barry Goldwater and Frank E. Moss to James C. Fletcher, as quoted in "Aircraft Fuel Conservation Technology Task Force Report," Office of Aeronautics and Space Technology, Sept. 10, 1975.

26. Bill Siuru and John D. Busick, *Future Flight: The Next Generation of Aircraft Technology* (Blue Ridge Summit, PA: TAB/AERO, 1994), p. 37.

27. Gerald G. Kayten, NASA Headquarters Director of Study and Analysis, to Lewis Research Center director, Dec. 7, 1973, Box 181, Division 8000, NASA Glenn archives.

Dr. James Fletcher appearing before the U.S. Senate Committee on Aeronautical and Space Sciences to discuss Skylab (May 23, 1973). As NASA Administrator, Fletcher gained the approval of the Nixon Administration to develop the Space Shuttle as a follow-on human space flight effort. (NASA Headquarters—Greatest Images of NASA [NASA HQ GRIN].)

American Airlines, would explore several opportunities for achieving more energy-efficient aircraft. When the two Senators challenged NASA's leaders in January 1975 to come up with a solution for the crisis threatening American aviation, they drew their inspiration from these programs.[28]

NASA's Administrator, James Fletcher, assigned overall responsibility for a new airline fuel efficiency program to Alan M. Lovelace, NASA's Associate Administrator for Aeronautics and Space Technology.

With a goal of conservation before him, in a month's time, Lovelace had established the Aircraft Fuel Conservation Technology Task Force. James J. Kramer, from the Office of Aeronautics and Space Technology

28. Ethell, *Fuel Economy in Aviation*, p. 7.

(OAST), directed the 15-member task force, which came to be called the Kramer Committee.[29] For the next 2 months, the Committee members worked together to develop a technology plan to satisfy the Government's request. To evaluate their results, NASA on April 17 established an advisory board chaired by Raymond L. Bisplinghoff from the University of Missouri.[30] The Kramer Committee included a remarkably diverse and knowledgeable group of members representing universities (MIT), industry (American Airlines, Pan American, Douglas Aircraft, Boeing), Government (NASA, the Federal Aviation Administration [FAA], the Department of Transportation, the Department of Defense), and engine manufacturers (Pratt & Whitney, Lockheed, General Electric). They named the new conservation effort the Aircraft Energy Efficiency (ACEE) program.

John Klineberg, one of the key members of the task force, recalled how closely it listened to the needs of industry. As part of the process, task force members went directly to industry leaders. They then reported areas of concern and need back to the task force. They discussedissues with the various NASA Centers in the same way. The information was then turned into briefings, and the task force communicated the results back to industry and NASA. This was the process by which ACEE took shape.[31]

In May and June 1975, the advisory board met to review and revise the initial recommendations of the Kramer Committee. The group members initially started with a long list of initiatives that would potentially lessen

---

29. The task force included: Jim Kramer (director), John Klineberg, Bill McGowan, Fred Povinelli, and Bill Roudebush, OAST; Darrell Wilcox and Lou Williams, Ames Research Center; Del Nagel, Joe Alford, and Dal Maddalon, Langley Research Center; Milt Beheim and Dick Weber, Lewis Research Center; Dick Baird, the Department of Defense; Bill Devereaux, the Department of Transportation; and Herm Rediess and Joe Tymczyszyn, Federal Aviation Administration.

30. The advisory board included: Raymond L. Bisplinghoff, University of Missouri; Jack L. Kerrebrock, the Massachusetts Institute of Technology; Franklin W. Kolk, American Airlines; John G. Borger, Pan American World Airways; Ronald Smelt, Lockheed Aircraft Corporation; Charles S. Glasgow, Jr., Douglas Aircraft Company; Abe Silverstein, a former NASA Center Director; Michael I. Yarymovych, the Energy Research and Development Administration; William E. Stoney, the Department of Transportation; Robert N. Parker, the Department of Defense; Richard Coar, Pratt & Whitney Aircraft Division; H.W. Withington, Boeing Commercial Airplane Company; and Edward Woll, General Electric.

31. Klineberg, interview with Bowles, July 28, 2008.

the effects of the energy crisis. They then worked to reduce the options to a manageable number and divided them into specific technology sections. Although some projects might be ready for short-term implementation, most required a projected 10 years of research and development before aircraft fuel consumption would be reduced. The ultimate goal, according to Kramer, was "achieving a technology readiness by 1985 for a 50 percent reduction in fuel consumption for new civil transports."[32] There were two unbreakable ground rules for attaining these goals. The first was that fuel would not be saved at the expense of the environment. The second was that fuel savings techniques would not compromise aircraft safety in any way. Ultimately, the Kramer Committee identified six technology plans it believed would achieve the stated fuel reduction goal, without violating safety or environmental criteria.

The Kramer Committee's six conservation technologies addressed the three ways to improve fuel efficiency in an airplane, as expressed in the Breguet range equation: decreasing the fuel consumption by an engine, decreasing the aircraft drag by improving its aerodynamics, and decreasing the weight of the airplane. The Committee's six conservation technologies addressed all of these areas. The Engine Component Improvement project would identify minor ways to improve existing engines to make them more fuel efficient. At an estimated cost of $40 million, the cumulative effects could have a 5-percent increase in fuel savings. The Energy Efficient Engine (or "E^3," as it came to be known) project would go beyond modifications to existing engines by creating an entirely new model to be ready for airplanes built in 1990. This was a planned 7-year, $175-million project, with a potential 10-percent fuel savings. The final propulsion project was considered the most radical of the all, a return to propellers, or "turboprops." Though the riskiest proposal in terms of success, it also provided one of the greatest rewards, a potential 15- to 30-percent fuel savings compared with existing jet aircraft. The turboprops were a 9-year, $125-million program.

In addition to the three propulsion projects, the Kramer Committee also identified two main airframe aerodynamics performance initiatives.

---

32. Statement by Kramer to the House of Representatives Subcommittee on Aviation and Transportation R&D Committee on Science and Technology, Sept. 1975, Box 179, Division 8000, NASA Glenn archives.

James Kramer, Associate Administrator for Aeronautics, visiting Langley in 1978. Joseph Chambers, at right, briefs him on stall/spin research for general aviation airplanes. Kramer is holding a spin tunnel model of the American Yankee airplane shown in the background. Courtesy of Joseph Chambers.

Dr. Hans Mark (1929–) speaks Moffett Field Officer's Club (November 9, 1976). Mark became NASA Deputy Administrator in July 1981. He had served as Secretary of the Air Force and as Undersecretary of the Air Force. Mark has also served as Director of NASA's Ames Research Center, Mountain View, CA. (NASA Ames Research Center [NASA ARC].)

The first, the Energy Efficient Transport program, called for evolutionary improvements to optimize aircraft designs. Wind tunnel studies would help verify new designs that decreased drag and improved fuel efficiency. This was a 7-year, $50-million program, with estimated fuel savings of 10 to 15 percent. A second aerodynamics initiative, Laminar Flow Control, had an even greater potential drag reduction potential through a smooth (or laminar) flow over the wings and tail. Virtually all civil transports cruise with a turbulent flow that increases drag. With anywhere from 20 to 40 percent fuel savings, the 10-year, $100-million program was estimated to be flight-ready by 1990.

The final area the Kramer Committee identified involved using advanced materials to reduce the weight of aircraft. The Composite Primary Aircraft Structures program investigated composites containing boron or graphite filaments in polyimide, epoxy, or aluminum matrices that could potentially reduce aircraft weight by 25 percent. This was a $180-million, 8-year program with 10- to 15-percent fuel savings potential, with the new composite designs in service by 1985.

There were some concerns about the selection of these ACEE projects. Hans Mark, the Director of NASA's Ames Research Center, wrote to Alan Lovelace in June 1975 saying that "Certainly there are many other aeronautical needs which must not be neglected."[33] He understood that fuel conservation was in the national interest, but he cautioned against committing too much aeronautical funding to the development of civil aviation at the expense of military aircraft technology. He added that aeronautical priorities change quickly. The main issue in 1968 was airport congestion. In 1970, aircraft noise was the central problem. By 1974, it was fuel conservation. Mark wanted to ensure that NASA did not overreact to something that might turn out to be a short-term problem. Furthermore, he suggested that fuel efficiency could be improved by working with the airlines to develop more fuel-efficient flight trajectories.

Lovelace appreciated Mark's concerns, but the ACEE plan went forward without any changes. In total, six recommendations made by the Kramer Committee cost a projected $670 million, with a 10-year timeframe for implementation. The percentage fuel savings for each project could not be added together because they did not all apply to the same type of aircraft.

---

33. Hans Mark to Alan Lovelace, June 4, 1975, Box 181, Division 8000, NASA Glenn archives.

However, when combined, they did reach the stated goal of 50 percent in total fuel reduction. Raymond Bisplinghoff, the head of the advisory board for the Kramer Committee, officially presented these conclusions and an outline of the technology plan to Alan M. Lovelace on July 30, 1975.[34]

The role of NASA itself was the one of the final areas of debate by the Kramer Committee. Since the Committee was made up of a cross section of individuals from different academic, industrial, and governmental organizations, there was a broad and vigorous discussion about NASA and the importance of Government-funded research. The Committee members realized that ACEE was unusual because it "in some instances goes further in the demonstration of civil technology improvements than has been NASA's traditional role." But the consensus was that this was necessary because of the "inability of industry to support these activities on their own." Specifically, the Kramer Committee stated, individual technologies such as the turboprop or laminar flow control would likely never be developed by industry because of their "high technical risks."[35] The Committee published its final report, "Aircraft Fuel Conservation Technology," in September 1975.

Concurrently with the publication of the report, a separate and independent study examined the costs and benefits of implementing these projects. NASA contracted with Ultrasystems, a California company that specialized in generating computerized economic models. Looking at a 10-year period, Ultrasystems used the Kramer Committee's $670-million cost estimate and compared it with a forecast of commercial aircraft fleet fuel consumption. While Ultrasystems conceded that the airline industry was in a state of flux and was often unpredictable, it tried to use some baseline assumptions to predict the near-term future. To lessen errors, Ultrasystems used proprietary data given to NASA's Ames Research Center from various aircraft manufacturers and the airline industry itself. It concluded that implementing the six ACEE programs advocated by the Kramer Committee would save the equivalent of 677,500 barrels of

34. Bisplinghoff to Lovelace, July 30, 1975, as found in, "Aircraft Fuel Conservation Technology Task Force Report," Office of Aeronautics and Space Technology, Sept. 10, 1975.
35. "Advisory Board Report of the Third Meeting on Aircraft Fuel Conservation Technology," Box 179, Division 8000, NASA Glenn archives.

oil each day. The future price of a barrel of oil determined the ultimate potential return on this investment. Again, Ultrasystems made some educated assumptions but concluded that for each dollar spent on the program, there would be anywhere from a return of $7.50 to $26 on the investment. The final assessment examined whether funding for these programs should come from private industry or the Government, and it concluded, "It is extremely unlikely that private industry could meet the expected capital requirements of the NASA program and, consequently, Federal support is necessary."[36]

The engineers had finished defining and laying out the program. The only other question to be answered was: Would Government approve the program and provide the funding for one of the largest coordinated fuel conservation projects ever attempted in the United States? To answer that question, the Senate held three hearings in fall 1975 and used the testimony to decide the program's future.

Conservation Testimony on the Hill

With a compelling technology plan and a positive cost analysis in place, the next step was to hold congressional hearings to determine whether the program would be funded. Numerous high-level industrial, academic, and governmental executives with intimate knowledge of the airlines industry came to Washington, DC, to submit their personal statements as to the significance of the energy crisis and the importance of the NASA conservation plan. The Senate Committee on Aeronautical and Space Sciences led the hearings and planned to make its final conclusions known in early 1976.[37] The testimony was important because it provided the opinions from a cross section of those invested in the success of the United States airlines industry. The executives used this opportunity to talk about the work of the Kramer Committee, the ACEE programs in general, and their concerns for the future of the airlines industry.

---

36. "Examination of the Costs, Benefits and Energy Conservation of the NASA Aircraft Fuel Conservation Technology Program," Report No. 8291-01, Nov. 15, 1975, pp. 12, 18, and 32, as found in, Box 948, Division 2700, NASA Glenn archives.

37. The Senate Committee on Aeronautical and Space Sciences included: Chairman Frank E. Moss (D-UT), Stuart Symington (D-MO), Barry Goldwater (D-AZ), John C. Stennis (D-MS), Pete V. Domenici (R-NM), Howard W. Cannon (D-NV), Paul Laxalt (R-NV), Wendell H. Ford (D-KY), Jake Garn (R-UT), and Dale Bumpers (D-AR).

The hearings began September 10, 1975. One of the first to speak was Clifton F. von Kann, a senior vice president of the Air Transport Association of America (ATA). This was an organization that represented nearly all of the individual carriers in United States airlines industry. He said that aviation was more than just a transportation system, because it was one of the main sources of American power. It played a central role in military strength and was a key component of the domestic economy. But he warned that since "fuel is the life blood of the airlines . . . [and] the airline industry is a basic building block of the U.S. economy," the energy crisis posed a serious threat to the Nation. If nothing were done to counteract it, this presented a danger for the entire United States airline system. Von Kann concluded by supporting the Kramer Committee and the NASA conservation program, saying, "The NASA program offers the prospect of major benefits to aviation economics as well as fuel conservation in itself. . . . We recommend approval."[38]

The military also offered supportive testimony on Capitol Hill. Walter B. LaBerge, Assistant Secretary of the Air Force Research and Development, provided the perspective of the military branch in which fuel conservation proved most vital. The Air Force used 50 percent of the fuel allocated to the Department of Defense, and it had representatives who worked with NASA on the Kramer Committee. LaBerge said, "We are enthusiastic about NASA's plan. . . . [It] will directly benefit the nation and the Department of Defense."[39]

NASA's leaders also provided testimony and made compelling arguments for approving the plan. George M. Low, Deputy Administrator, said that 78 percent of aircraft flying in the Western World were manufactured in the United States. Without a concentrated effort to develop fuel-efficient aircraft, this $4.7-billion export industry would evaporate. Low said, "Our world leadership in aviation is in serious danger today."[40] Alan Lovelace, NASA's Associate Administrator, alluded to advances by the Soviet Union

---

38. Statement by Clifton F. von Kann to the Senate Committee on Aeronautical and Space Sciences, Sept. 10, 1975, Box 179, Division 8000, NASA Glenn archives.

39. Statement by Walter B. LaBerge to the Senate Committee on Aeronautical and Space Sciences, Sept. 10, 1975, Box 179, Division 8000, NASA Glenn archives.

40. Statement by George M. Low to the Senate Committee on Aeronautical and Space Sciences, Sept. 10, 1975, Box 179, Division 8000, NASA Glenn archives.

as an incentive to improve fuel efficiency. He indicated that the Russians had already developed a high-efficiency turboprop that could cruise at speeds near those of existing jets and that the United States should provide the resources to keep pace.[41] Raymond Bisplinghoff, Chairman of the Kramer Committee's Advisory Board, simply said that funds allocated to the development of aircraft conservation technology "was a better investment than the continued importation of middle-eastern oil."[42]

A second day of testimony took place October 23, 1975. George H. Pedersen, a technical coordinator for the American Institute of Aeronautics and Astronautics (AIAA), presented the findings of the "AIAA technical hierarchy." All of the members, he said, "strongly endorsed" and gave "universal approval" to all the proposals by the Kramer Committee. They agreed that industry alone could never achieve a program of this magnitude by itself. Capital risk aversion in the airline industry prevented it from investing in long-term fuel conservation technology. Only NASA could achieve this, in their opinion. The AIAA's only negative criticism was that alternative fuels were not a component of the conservation program.[43]

Karl G. Harr, Jr., the president of the Aerospace Industries Association, added to the positive assessments made by the AIAA. He noted that while NASA would take the lead in the project, industry would also play a vital role in its successful outcome. The Kramer Committee envisioned the program as a joint effort between Government and industry, and Harr agreed that while the research took place within NASA, the certification and production phases should be the responsibility of industry. This would be the best way to assure rapid technological development and technology transfer to industry.[44]

Industry representatives, including Boeing, Pratt & Whitney, and General Electric, all voiced enthusiasm for the program. A Boeing vice president stated in his testimony that "the NASA research program should, in the long run, result in major U.S. fuel saving [and] preservation of U.S.

---

41. Statement by Alan M. Lovelace to the Senate Committee on Aeronautical and Space Sciences, Sept. 10, 1975, Box 179, Division 8000, NASA Glenn archives.

42. Statement by Raymond L. Bisplinghoff to the Senate Committee on Aeronautical and Space Sciences, Sept. 10, 1975, Box 179, Division 8000, NASA Glenn archives.

43. Statement by George H. Pedersen to the Senate Committee on Aeronautical and Space Sciences, Oct. 23, 1975, Box 179, Division 8000, NASA Glenn archives.

44. Statement by Karl G. Harr, Jr., to the Senate Committee on Aeronautical and Space Sciences, Oct. 23, 1975, Box 179, Division 8000, NASA Glenn archives.

technological leadership."[45] Likewise, a Pratt & Whitney vice president said that from his perspective as an aircraft engine manufacturer, "There is no doubt in my mind that the implementation of NASA's plan can play a significant role in achieving the National fuel conservation goals for our air transport system."[46] A General Electric vice president testified that his company "strongly endorse[d]" the NASA program; quite simply, he explained, it would help the United States reach "energy independence."[47]

The Senate Committee held its final day of testimony November 4, 1975. Offering an international perspective was Yuji Sawa, a vice president from All Nippon Airways, which was the largest airline in Japan and the seventh largest in the world. According to Sawa, 80 percent of all Japan's fuel was imported from the Middle East, and as a result, the embargo stemming from the October 1973 crisis threatened his country and inspired drastic conservation measures. He concluded that if the United States developed a new generation of fuel-efficient aircraft, there would be a tremendous demand in Japan to start importing them to replace the existing fleet.[48] After Sawa spoke, the president of United Airlines, Charles F. McErlean, presented his views of the NASA program. With 49,000 employees and a fleet consuming 1.5 billion gallons of fuel annually, few other organizations had such a direct stake in the development of conservation technologies. He said that if the current energy trends continued, it would "jeopardize our future," explaining that, "This is not only a serious problem, it is a potentially crippling one."[49] McErlean concluded by endorsing the NASA conservation plan as one important measure to bring the airlines out of their fuel crisis.

Not everyone who testified on Capitol Hill was as positive. From a technological perspective, the NASA fuel conservation program appeared

---

45. Statement by John E. Steiner to the Senate Committee on Aeronautical and Space Sciences, Oct. 23, 1975, Box 179, Division 8000, NASA Glenn archives.

46. Statement by Bruce N. Torell to the Senate Committee on Aeronautical and Space Sciences, Oct. 23, 1975, Box 179, Division 8000, NASA Glenn archives.

47. Letter from Gerhard Neuman to the United States Senate Committee on Aeronautical and Space Sciences, Oct. 22, 1975, Box 179, Division 8000, NASA Glenn archives.

48. Statement by Yuji Sawa to the Senate Committee on Aeronautical and Space Sciences, Nov. 4, 1975, Box 179, Division 8000, NASA Glenn archives.

49. Statement by Charles F. McErlean to the Senate Committee on Aeronautical and Space Sciences, Nov. 4, 1975, Box 179, Division 8000, NASA Glenn archives.

sound and had tremendous industrial support. However, technologies are not developed in a vacuum. They are created and function within an environment that can be as important as the technology itself to overall success. Of these external forces, the economics of aircraft fuel conservation technology was the most vital, and some criticized the Kramer Committee and NASA for ignoring these economic issues. Representatives from the Federal Energy Administration and the Energy Research and Development Administration (ERDA) raised these concerns when they spoke before the Senate Committee.

Roger W. Sant, an Assistant Administrator for the Federal Energy Administration, was the first to raise a red flag about the program and voice a concern that he thought was a significant oversight. The problem, from his perspective, was that external factors were not analyzed carefully enough. "The report," he explained, "does not address several important issues which are critical to an understanding of the ultimate worth or merit of the proposed NASA research efforts." Namely, even if the technology were successfully developed, would economic factors be similar enough in 10 years to result in market demand for the new aircraft? Would the expense of building these energy-efficient planes be cost effective, and would the airlines be willing to make capital investments in them? "NASA apparently was not requested to undertake any such analysis of external variables," Sant said, concluding that even if NASA achieved its technological goals, "it is not clear that this research program represents the most cost-effective use of limited Federal Energy Research funds."[50]

Another voice of disapproval came from the Energy Research and Development Administration. Though it had a member on the Kramer Committee, during the testimony James S. Kane, Deputy Administrator, raised serious concerns. Although he stated that NASA had identified key areas of technological development that would be a positive force in airline fuel conservation, he believed the project's scope was too large. Kane said, "ERDA considers that NASA is asking for a disproportionate share of the money available for energy R&D." Furthermore, Kane thought the United States should not put such a large emphasis on airline fuel to the neglect of automobile fuel conservation. He presented compelling statistics to support his argument. The most recent consumption statistics showed

---

50. Statement by Roger W. Sant to the Senate Committee on Aeronautical and Space Sciences, Nov. 4, 1975, Box 179, Division 8000, NASA Glenn archives.

that automobiles made up more than 50 percent of the total petroleum usage, followed by trucks and buses at 20 percent. Although the airlines were consuming 11-billion gallons of fuel per year, this only represented 7 to 10 percent of the total national petroleum consumption. With limited research and development funds available, Kane questioned the wisdom of NASA's request of $670 million. He concluded, "The automobile should be supported as first priority with a much larger share of the budget."[51]

A third, less-well-known organization, ECON, Inc., of Princeton, NJ, presented its economic assessment of the NASA program. Though it was more supportive than ERDA or the Federal Energy Administration, it did present some important economic warnings. The fuel savings costs were unquestioned. Based upon a 3-month study, it concluded that the results of the new technologies would save 90.3 billion gallons of fuel or 2.15 billion barrels between 1976 and 2005. There were several economic factors that could potentially alter these savings. The most significant was whether the airlines would adopt the new technology, and how quickly the industry would render the old jets obsolete. Several factors dictated the replacement policy. These were the price and availability of fuel and the return on investment for the purchase and implementation of the new airplanes. If fuel constituted 20 to 35 percent of an airline's operating costs, ECON warned "the most profitable price for aircraft may not embody the technology for minimum fuel consumption."[52] Therefore, ECON recommended that the Government seriously consider offering economic incentives to encourage industry to adopt the new NASA technology when it is ready. Otherwise, its analysts feared, engineers might have successful technologies sitting unused because future fuel costs might not make implementation economically cost efficient.

After the final testimony, the hearings ended. NASA had completed the preliminary blueprint for the six conservation initiatives and could now only wait for Congress to decide what to do. Overall, one of the participants said, the hearings went well, and the "Senate audience was friendly."[53]

---

51. Statement by James S. Kane to the Senate Committee on Aeronautical and Space Sciences, Nov. 4, 1975, Box 179, Division 8000, NASA Glenn archives.
52. Statement by Klaus P. Heiss, president of ECON, Inc., to the Senate Committee on Aeronautical and Space Sciences, Nov. 4, 1975, Box 179, Division 8000, NASA Glenn archives.
53. Letter from James F. Dugan to the Director of Aeronautics, Sept. 15, 1975, as found in Box 179, Division 8000, NASA Glenn archives.

The Senators seemed receptive to the proposals and understood the severity of the situation confronting the United States. But the question remained: Would the Government approve the program? The answer came 3 months later.

ACEE Approval

On February 17, 1976, the 10 members of the Senate Committee on Aeronautical and Space Sciences published its report on the "Aircraft Fuel Efficiency Program." Frank Moss and Barry Goldwater, the Senators who wrote the original letter to the NASA Administrator about the need for a conservation solution to the aviation fuel crisis a year earlier, led the Committee. NASA's six-part response, coupled with broad support from industry, made a compelling case for funding the $670-million program, in the Committee's opinion.

The Moss and Goldwater Senate Committee published several conclusions regarding the hearings. Its members stated first that they had learned that fuel efficiency would play a cutting edge role in competing in the world aircraft market. Second, they believed that embarking on a fuel-efficiency program would serve as an important stimulus to the U.S. aircraft industry. The benefits would accrue not only to the traditional aircraft manufacturers and operators, but also to the numerous subcontractors. In 1974, more than half a million people were employed in the aircraft and parts industry. Third, new fuel-efficient aircraft would offer a major assistance to the entire air transportation system, which was struggling for profitability and survival in the midst of a fuel crisis and escalating oil prices. Although the technologies identified by NASA would not have an immediate impact, Moss and Goldwater concluded that "higher fuel costs will remain an urgent program in the foreseeable future . . . more fuel efficient aircraft will be highly desirable and beneficial to the air transportation system in the next and succeeding decades."[54]

Fourth, the Committee stressed that the project was important because it involved energy conservation. With the potential to reduce fuel consumption by up to 50 percent, its effects would include higher profits and more environmentally friendly technology, which included aircraft

54. "Aircraft Fuel Efficiency Program," Report of the Committee on Aeronautical and Space Sciences of the United States Senate, Feb. 17, 1976, p. 6.

noise and pollution reduction. By using less fuel, the market demand for oil would also decrease. Basic economics suggested that a reduction in demand would decrease prices, providing yet another way for airlines to increase profitability. Finally, in an area that was unquantifiable, yet perhaps undeniably the most important of all, the NASA program would strengthen the United States. Decreasing demand for fuel would "reduce the vulnerability of the . . . Nation to the whims of oil rich nations."[55] The technology could also be adapted by the military, with the Air Force incorporating fuel-efficient technology to increase the range of its bombers.

One other major conclusion of Moss and Goldwater's report was that it was the Government's responsibility to bear the risk and the costs of the technological research and development. There were several reasons for this. First, the Senators acknowledged the "considerable risk" associated with the projects. With the cost to develop a new airplane already at $1 billion, Moss and Goldwater said, "it is no mystery why aircraft manufacturers must be conservative in their choice of their technology."[56] Even if industry were to try to develop it, the technology would be proprietary, and the benefit to the Nation would be significantly reduced. So the Senators concluded that the aircraft fuel-efficiency program was a "classic example" of the need for Government support.

NASA was the appropriate Government Agency to take the lead, thanks to what the Senators recognized as its "long history of excellence." Although NASA had recently become associated with space exploration, they pointed out that "aeronautics is a part of the very name of the agency."[57] Also, very clearly, NASA was not in business to build airplanes—so though it would take the lead in the research, it would also determine the most opportune time to transfer the technology to industry.

Moss and Goldwater carefully took into consideration the criticisms of the program, most specifically those of the Federal Energy Administration. In addressing the first argument, that no one really knew how much fuel savings these program would create, the Senators responded that this was precisely the reason for the research project in the first place—so that NASA could determine which projects were feasible. As for whether the

55. Ibid., p. 7.
56. Ibid., p. 9.
57. Ibid., p. 10.

airlines would not have the capital to purchase the new technological efficient airplanes in 1985, the Senators were confident that the aging air transportation system, which would be ready for replacement with some type of aircraft, would offer a fuel-efficient solution to the problem Why would airlines executives purchase the older, less-fuel-efficient aircraft? As to the third argument, the question was: Why focus on aircraft when automobiles consumed so much more fuel? The Senators responded to concerns about the program's focus on aircraft over automobiles by explaining that no automotive fuel conservation program was being proposed, and that the airline fuel-efficiency program should be judged on its own merits.

The Senators agreed that it was impossible to predict the real industry costs for implementing this advanced technology, recognizing that the proposed program's only established costs were the amounts required for NASA's research. On the other hand, they argued that there was no indication that the costs would be prohibitive, and the fact that there was broad approval from industry supported this assumption. In rejecting the notion that all costs should be established up front, the Senators wrote, "We believe it would be a mistake to insist that all the costs (or benefits) must be assessed before a decision on the program can be made." They justified being lax in their demands for financial analysis because of their unquestioned belief in the importance of the fuel-efficient technology. Furthermore, it is almost impossible to perform a cost-benefit analysis on basic research. On this point, they simply responded, "We feel this program has the potential of returning enormous benefits, far in excess of the costs."

In conclusion, the Senators approved the NASA plan. "On the basis of fuel savings alone," they wrote, "the aircraft fuel efficiency program is attractive. And considering all the potential benefits and NASA's mandate to maintain U.S. leadership in aeronautics, the program is essential."[58] With its mandate in place, NASA immediately went to work to put the plan into action and divided responsibility for the six technology initiatives between two NASA Centers. It would fall upon the shoulders of Lewis Research Center in Ohio and Langley Research Center in Virginia to make these technological dreams a reality.

---

58. Ibid., p. 13.

Chapter 2:

Threads and Sails at Langley

Several revolutionary advances in aircraft design and technology emerged from NASA's Langley Research Center in the 1970s as a result of the Aircraft Energy Efficiency program. Historian Roger Bilstein referred to them as "arcane subjects" that included "some unusual hardware development."[1] The first of them challenged the seemingly obvious assumption that the aircraft should be made of metal and aluminum. Most people took for granted that these were the best aircraft materials until a 1978 *Los Angeles Times* reporter speculated that "Large commercial and military jets of the future will probably be made not of metal but of thread."[2] Machines with "gigantic spools of yarn" began making airplane parts with "threads" from new composite materials that promised tremendous weight savings, thereby making airplanes more fuel efficient.[3]

Another Langley development fundamentally changed the shape of the aircraft in two key ways. The first idea was an airplane wing that emulated a boat "sail" through a change in its shape at the wingtip. These first became visible in the 1980s, when the main wings on some aircraft, including the MD-11 in 1986, took an unusual upward or vertical extension at the end of the wing. This acted like a sail, taking advantage of a whirlpool of air that naturally occurred around the wingtip. The sail caught the swirling air, transformed it into forward thrust, reduced drag, and increased fuel efficiency.[4]

---

1. Roger E. Bilstein, *Testing Aircraft, Exploring Space* (Baltimore: The Johns Hopkins University Press, 2003), p. 123.
2. Charles Hillinger, "Jet Fighter Made of Thread," *Los Angeles Times*, Nov. 6, 1978.
3. Anthony Ramirez, "Advanced Composite Construction," *Los Angeles Times,* Sept. 4, 1984.
4. James J. Haggerty, "Winglets for the Airlines," *Spinoff 1994,* (Washington, DC: NASA, 1994), pp. 90–91.

A second development, called the "supercritical wing," was so unusual that when it emerged from a Langley wind tunnel, even its designer confessed, "nobody's going to touch it with a ten-foot pole without somebody going out and flying it."[5]

These striking new technological developments—threaded aircraft materials, wing "sails," and supercritical wings—were the primary focus of two ACEE projects led by Langley engineers. They were airframe technology advances with the primary goals of reducing structural weight and improving aerodynamic efficiency as a means of decreasing fuel consumption.[6] The threaded aircraft materials were part of the Composite Primary Aircraft Structures (CPAS) program. The sailboat emulation, officially known as a "winglet," and the supercritical wing were two of the most successful components of the multifaceted Energy Efficient Transport (EET) program. A central portion of Langley's contribution to the ACEE program, these projects achieved significant savings in fuel economy.[7]

The Flying Field—Langley Research Center

In his autobiographical novel, *Look Homeward, Angel*, Thomas Wolfe described summer 1918, when, as a young man, he went looking for work in Hampton, VA. There, at a place called the "Flying Field," he observed gangs of workers engaged in "grading, leveling, blasting from the spongy earth the ragged stumps of trees and filling interminably, ceaselessly, like the weary and fruitless labor of a nightmare, the marshy earth-craters, which drank their shoveled toil without end."[8]

---

5. Richard Whitcomb, quoted in Tom D. Crouch, *Wings: A History of Aviation from Kites to the Space Age* (New York: W.W. Norton & Co., 2003), p. 462.
6. Robert W. Leonard and Richard D. Wagner, "Airframe Technology for Energy Efficient Transport Aircraft," *Aerospace Engineering and Manufacturing Meeting of the Society of Automotive Engineers, Nov. 29–Dec. 2, 1976.*
7. James Schultz, *Winds of Change: Expanding the Frontiers of Flight: Langley Research Center's 75 Years of Accomplishment 1917–1992* (Washington, DC: NASA, 1992). Roger D. Launius and Janet R. Daly Bednarek, eds., *Reconsidering a Century of Flight*, (Chapel Hill: University of North Carolina Press, 2003). Tom D. Crouch, *Wings: A History of Aviation from Kites to the Space Age* (Washington, DC: Smithsonian National Air and Space Museum, 2003).
8. Thomas Wolfe, *Look Homeward, Angel: A Story of the Buried Life* (New York: Charles Scribner, 1957), p. 121.

Langley Memorial Aeronautical Laboratory in the 1920s. One enduring feature was the mud around the administration building. (NASA Langley Research Center [NASA LaRC].)

Wolfe's evocative prose spoke of the human muscle required to construct a facility devoted to escaping the bounds of Earth. What these men achieved was the construction of the only American civilian aviation laboratory until 1941. The laboratory became the first Center of the newly created National Advisory Committee for Aeronautics (NACA).

The "Flying Field" was named for aviation pioneer Samuel P. Langley, a Harvard University professor of astronomy and Secretary of the Smithsonian Institution.

In the 1890s, he became obsessed with flying aircraft, but his unusual "aerodrome" experiments resulted in spectacular crashes, and the press began referring to his machines as "Langley's folly."[9]

He died in 1906, having never flown, but his namesake laboratory would become one of the leading centers of aeronautical research in the world. The NACA charter of 1915 defined a very specific mission: to "supervise and direct the scientific study of the problems of flight with a view to their practical solution." This practical emphasis meant that Langley's

9. "Poor Old Langley," *Los Angeles Times*, May 27, 1918.

Samuel Pierpont Langley (1834–1906) and Charles M. Manley, left, chief mechanic and pilot onboard the houseboat that served to launch Langley's aerodrome aircraft over the Potomac River in 1903. (NASA Langley Research Center [NASA LaRC].)

The Langley aerodrome (December 8, 1903). After this photo was taken, the project ended in failure when it fell into the Potomac River. (NASA Langley Research Center [NASA LaRC].)

engineers would treat aeronautical problems not from a theoretical distance, but through the reality of actual aircraft in flight. More than any other American institution, it was responsible for the research necessary to solve the problems of flight and develop the airplane into both a commercial product and a centerpiece of the Nation's defense.[10] Commissioned in 1920, its early years were filled with both promise and hardship. Langley's three original buildings included a wind tunnel, an engine-dynamometer laboratory, and a research laboratory that became what some described as an "aeronautical mecca" in the United States.[11]

The earliest aeronautical work at Langley included the construction and use of experimental wind tunnels, the first in 1920, to test new aircraft designs. Because the wooden biplanes of the 1920s were so frail, engineers

10. "Langley Research Center, Research Highlights, 1917–1967," Unit 4A, Cabinet 4, Shelf 3, Langley History Box 1, NASA Langley archives.

11. Alex Roland, *Model Research: The National Advisory Committee for Aeronautics, 1915–1958,* vol. 1 (Washington, DC: NASA SP-4103, 1985), p. 83. Hansen, *Engineer in Charge,* pp. xxxi–xxxii.

Langley Laboratory's first wind tunnel, a replica of a 10-year-old British design, became operational in June 1920. (NASA Headquarters—Greatest Images of NASA [NASA HQ GRIN].)

had a tremendous opportunity to improve aerodynamic efficiency through their research. They first began asking questions about how the shape of wings would decrease drag, how to design propellers, when to best use flaps, and how to predict control forces on various aircraft components.

Langley's engineers developed the world's first full-scale research tunnel for propellers in 1926, and its work in drag reduction and retractable landing gear were among some of its first major technical breakthroughs. The engineers also developed the "NACA cowling," which covered the engine, significantly reduced drag, and improved engine cooling. While all these advances required fundamental research, the ultimate goal was the practical application.

Practical achievements continued for the next several decades. In the 1930s, Langley's laboratory tests contributed to the development of advanced aircraft such as the Douglas DC-3 and the Boeing B-17. During World War II, Langley's engineers worked to improve the performance capabilities of military aircraft. In 1944, the NACA was in the process of testing 78 different types of aircraft, and a vast majority of these tests were done at Langley. But by this time, Langley was no longer the NACA's only Center. The NACA established Ames Research Center in California to complement Langley in 1940. One year later, the NACA opened Lewis Research Center in Cleveland, OH, to focus on engine propulsion. After the war, Langley's engineers explored the unknown areas of supersonic flight with jet aircraft (the "X" series of experimental aircraft) as well as vertical take-off and landing helicopters. But times were changing. Langley was no longer the sole NACA Center, and the American aeronautical landscape was a much different place.

After World War I, the NACA had a clear-cut vision: improve American aeronautics. After World War II, it struggled to find its way. During the war, jet propulsion emerged, and many hoped that the NACA and Langley would take the lead in probing the frontiers of this new revolution in flight. But the NACA now had competition. The U.S. Air Force had grown to become a branch of the military, with equal status to the Army, Navy, and Marines, and it began conducting its own aeronautical research and development. At the same time, aircraft-manufacturing became the largest industry in the United States. Not only was it also capable of its own research, but it depleted some of the NACA's talent pool by luring the best young aeronautical engineers with far better paying positions than the NACA could afford. This power and research potential gave industry a much stronger voice in dictating the direction and pace of research. The NACA needed to stake out its own sphere of influence in the postwar world, but its aging engineers were increasingly responding to the demands of the United States aircraft industry.[12]

The NACA needed revitalization, but this was not to be. Alex Roland described the 1950s as a time when the NACA seemed to be "waiting for the match." Other historians, including Virginia Dawson and James Hansen, have demonstrated that the NACA was still making contributions in the 1950s, among them axial compressors and supersonics, but in many respects, as Dawson suggested, "The difference was that the air force was now calling the shots."[13] Hansen also described the 1950s as an important time of transition in aeronautics. He wrote, "As the golden age of atmospheric flight

12. Roland, *Model Research,* p. 225.

13. Dawson, correspondence with Bowles, May 3, 2009. Dawson, *Engines and Innovation: Lewis Laboratory and American Propulsion Technology* (Washington, DC: NASA SP-4306, 1991), pp. 163–166.

reached full maturity in the 1950s—with only a few major things (like supersonic transport) left undone—many [engineers] . . . moved successfully from their mature aeronautical specialties into the new ones of spaceflight and reentry."[14]

As Roland characterized it, in 1957 Sputnik "provided the spark that set it off and . . . soon the old agency was consumed in flames."[15]

Although NASA, with an emphasis on space, replaced the NACA in 1958, aeronautics remained an important component of the new Agency, and aviation research continued at Langley.[16] But aviation was no longer an "infant technology." The NACA had achieved much, and the military and industry were also engaging in their own research and building their own test facilities. So aeronautics in the newly formed NASA often took a back seat to the more visible successes of the Apollo program. It maintained some of its greatest practical aviation importance and vitality, however, through service to the aircraft industry, which still needed the support that only Government could provide in leading-edge technology. This was best exemplified by the ACEE project, and Langley took a leading role.

By the 1970s, the aircraft industry in the United States was extremely important to the economic health of the Nation, and it made up a significant percentage of its positive balance of trade, second only to agriculture. International sales of American-manufactured aircraft from 1970 to 1975 totaled $21 billion. Robert Leonard, Langley's ACEE Project Manager, said that the export of a single jumbo jet equaled the importation of 9,000 automobiles.[17] However, this dominance was not assured. In 1978, Ralph Muraca, Langley's Deputy ACEE Project Manager, said there was a "real threat" to United States' dominance after other nations began developing new, efficient planes. Muraca concluded, "Clearly the importance of capture of most of this large market segment by our industry cannot be underestimated."[18] Just because the United States held onto this market in the mid-1970s did not mean its dominance would last. This was

---

14. Hansen, *Engineer in Charge,* p. 396.
15. Roland, *Model Research,* p. 283.
16. Chambers, *Innovation in Flight*.
17. Robert W. Leonard, "Fuel Efficiency Through New Airframe Technology," NASA TM-84548, Aug. 1982.
18. Albert L. Braslow and Ralph J. Muraca, "A Perspective of Laminar-Flow Control," *AIAA Conference on Air Transportation, Aug. 21–24, 1978*, p. 2.

especially true if it failed to develop fuel-efficient aircraft. As the price of jet fuel increased, fuel-efficient aircraft became move coveted throughout the world. Craig Covault, from *Aviation Week & Space Technology*, simply referred to this as "the challenge."

One key challenger was Airbus. Airbus began in the mid-1960s as a consortium of European aviation firms, and its mission was to compete directly with the American-dominated industry. In 1967, the first A300 appeared—a 320-seat, twin-engine airliner. In the late 1970s, Langley managers used a picture of a new French Airbus draped in Eastern Air Lines colors to illustrate the European threat. Donald Hearth, the Langley Director, said that because of this competition, his Center would begin restricting the flow of research results derived from the ACEE program to Europe. He said, "It is going to present an awkward situation and a change in the way we operate, and I'm not quite sure what it all means yet."[19] One thing was certain—ACEE was the most vital aeronautics program in the United States. Not only did it shoulder the burden and expectation of freeing the airline industry from the effects of the energy crisis, but the ACEE programs became the chief strategic hope to ensure American-made dominance of next-generation aircraft in the world's skies. The importance of ACEE, Leonard said, "cannot be overstated."[20] One of the more vital ACEE initiatives was research focusing on the materials used in the manufacture of airplanes, which probed the potential not for stronger or less expensive materials, but lighter ones.

A Strategic Center of Gravity—Composite Materials for Aircraft

Since the beginning of aviation history, weight reduction has been a primary goal.[21] During the time of the Wright brothers' first flights, airplanes were constructed of various types of wood, fabric, and wires.[22] It was

19. Hearth, quoted in Craig Covault, "Langley Aiding Transport Competition," *Aviation Week & Space Technology*, Aug. 21, 1978, p. 14.
20. Leonard, "Fuel Efficiency Through New Airframe Technology," NASA TM-84548, Aug. 1982.
21. Chambers, *Concept to Reality: Contributions of the NASA Langley Research Center to U.S. Civil Aircraft of the 1990s* (Washington, DC: NASA SP-2003-4529, 2003).
22. Tom Crouch, *The Bishop's Boys: A Life of Wilbur and Orville Wright* (New York: W.W. Norton & Co., 1989).

not until the 1920s that one of the first materials breakthroughs occurred, the Ford Tri-motor, dubbed the "Tin Goose." Henry Ford began manufacturing these aircraft in 1925, and they were unusual because of their use of metal and aluminum. The first planes used a corrugated metal shell, which surrounded a metal truss framework. In the 1930s, stressed-skin aluminum monocoque construction techniques emerged, and Langley played a key role in developing stress and strength analyses of the materials. These analyses paved the way for other structural and materials advances at Langley, which included thin wings for military aircraft in the 1940s. The new aircraft were required to withstand the stresses resulting from much faster speeds and also greater dynamic loads and vibrations. Langley engineers helped to pioneer the use of higher-strength alloys that prevented the aircraft from breaking apart under these forces. In the 1960s, a new type of material emerged that would come to challenge the dominance of metals in the skies. These were known as composites.

In 1967, at the 50th anniversary of the birth of the Langley laboratory, engineers announced that they were on the verge of several revolutionary new aircraft concepts, one of which was in materials.[23] The size of aircraft hadincreased dramatically since the time of the first airplanes. The Wright brothers' historic first aircraft was a fragile device that weighed just 1,260 pounds. In comparison, the all-metal 747 aircraft, which first flew 1 year after this Langley celebration, weighed 750,000 pounds. The fuel required to lift and propel these massive, metallic beasts was immense, so any weight reduction achieved through new materials was eagerly anticipated. Langley engineers believed they were on the cusp of achieving a major advance in composites.[24]

When two or more substances are combined together in one structure, the resulting material is called a composite. Aircraft composites are made by bonding together a primary material that has strong fibers with an adhesive, such as a polymer resin or matrix. These are various types of graphite, glass, or other synthetic materials that can be bonded together in a polymer epoxy matrix. The composite materials are typically thin-thread cloth layers or flat tapes that can be shaped into complex and

---

23. "Aeronautics—A Century of Progress," Presented at the *50th Anniversary Celebration and Inspection of Langley Research Center, Oct. 2–6, 1967*, Unit 4A, Cabinet 4, Shelf 3, Langley History Box 1, NASA Langley archives.
24. Ethell, *Fuel Economy in Aviation*, p. 59.

The corrugated shell is made from thermoplastic composite materials (February 17, 1978). (NASA Langley Research Center [NASA LaRC].)

aerodynamically smooth shapes of virtually any size.[25] Their application to aircraft led some to imagine the "Jet Fighter Made of Thread."[26] The physical properties of these materials made them extremely attractive in aircraft design because they were stronger, stiffer, and lighter than their metallic counterparts. Composites were also resistant to corrosion, a constant plague on metal aircraft. While efforts to incorporate these materials had been ongoing for several years prior to the 1970s, there were difficult hurdles that prevented their adoption. First was the general uncertainty as to whether they would actually work and could withstand the rigors of flight. The second was the cost of research and development simply to reach the stage at which they could be flight-tested. The cost of fabrication for production applications was, and still is, a key factor. Finally, there were no

---

25. John Cutler and Jeremy Liber, *Understanding Aircraft Design,* 4th ed. (Oxford: Blackwell Publishing, 2005), p. 159.
26. Charles Hillinger, "Jet Fighter Made of Thread," *Los Angeles Times,* Nov. 6, 1978.

data on their durability and maintenance requirements over time. As one observer stated, "The planned application of composites would require the development of revolutionary technology in aircraft structures."[27]

This development became the focus of 1972 joint Air Force-NASA program known as Long Range Planning Study for Composites (RECAST). The success of these investigations led NASA to include it as one of the six main program elements of ACEE, and it became known as the Composite Primary Aircraft Structures. Langley Research Center was to coordinate the program in conjunction with its industry partners: Boeing Commercial Airplane, Douglas Aircraft, and Lockheed. Langley was the obvious choice for this program, because the Center had played a leading role for decades in investigating aircraft structures and materials. The stated objective of CPAS was to "provide the technology and confidence for commercial transport manufacturers to commit to production of composites in future aircraft."[28] The technology included the development of design concepts and the establishment of cost-efficient manufacturing processes. The confidence would come with proof of the composite's durability, cost verification, FAA certification, and ultimately its acceptance by the airlines.

The main goal was to reduce the weight of aircraft by 25 percent through the use of these new materials, thereby decreasing fuel usage by 10 to 15 percent. Using composites for the wings and fuselage promised the greatest savings, but this was also the most technically challenging because these components were so vital to aircraft safety. To overcome some of the uncertainties of the materials, secondary structures (upper aft rudders, inboard aileron, and elevators) were the first candidates for composite materials.

Once these investigations were successful, then the development of medium-size primary structures (vertical stabilizer, vertical fin, horizontal stabilizer) would begin. In the meantime, some preliminary wing work would be explored, followed by work on the fuselage. Louis F. Vosteen headed the program at Langley.[29]

---

27. Marvin B. Dow, "The ACEE Program and Basic Composites Research at Langley Research Center (1975 to 1986)" (Washington, DC: NASA RP-1177, 1987), p. 1.
28. "ACEE Program Overview," NASA RP-79-3246(1), July 31, 1979, Box 239, Division 8000, NASA Glenn archives.
29. L.F. Vosteen, *Composite Structures for Commercial Transport Aircraft* (Washington, DC: NASA TM-78730, June 1978).

Composite elevators in flight evaluations on Boeing 727 during ACEE program. Courtesy of Joseph Chambers.

Secondary structures are those that have light loads and are not critical to the safety of the aircraft. The upper aft rudder on the Douglas DC-10 was one of the first of the secondary structures to be studied.[30] The rudder is a movable vertical surface on the rear of the vertical tail and is used for coordinating turning maneuvers and trimming the aircraft following the loss of an engine. Work to construct composite upper aft rudders actually began in 1974 but was completed as part of the ACEE program. Twelve units were put into service, and ACEE engineers estimated that manufacturing would cost less than metal after 50 to 100 units were installed. These units resulted in a 26.4-percent weight savings over the traditional aluminum alloy previously used for the rudder. Elevators were the next secondary structural components designed. Located at the rear of the fixed horizontal surfaces, elevators are movable surfaces used for controlling the longitudinal attitude of the airplane. Ten units were designed for the Boeing 727, and flight-testing began in March 1980.[31]

---

30. A. Cominsky, *Manufacturing Development of DC-10 Advanced Rudder* (Washington, DC: NASA CR-159060, Aug. 1979).
31. D.V. Chovil, S.T. Harvey, J.E. McCarty, O.E. Desper, E.S. Jamison, and H. Snyder, *Advanced Composite Elevator for Boeing 727 Aircraft,* vol. 1 (Washington, DC: NASA CR-3290, Nov. 1981).

Composites technology was applied to other projects as well. The Rutan Model 33 VariEze was built by the Model and Composites Section of Langley and then tested in a tunnel. (July 17, 1981). (NASA Headquarters—Greatest Images of NASA [NASA HQ GRIN].)

With a 23.6-percent decrease in the plane's weight, Boeing considered the production a success and approved the elevators for use on the 757 and 767. The final secondary structures were the inboard ailerons, moveable surfaces located on the edges of the wing.[32] Working in conjunction

32. C.F. Griffin, L.D. Fogg, and E.G. Dunning, *Advanced Composite Aileron for L-1011 Transport Aircraft: Design and Analysis* (Washington, DC: NASA CR-165635, Apr. 1981).

NASA's Boeing 737 in front of the hangar after its arrival in July 1973. Much ACEE work was performed on the 737 in later years. (NASA Langley Research Center [NASA LaRC].)

with the rudder, in-board ailerons are used for banking the airplane during high-speed turning maneuvers. Installed on a Lockheed L-1011 airplane, eight units began flight-testing in 1982. These were a significant improvement over the aluminum ailerons, reducing the weight by 65 pounds, the number of ribs from 18 to 10, and the number of fasteners from 5,253 to 2,564.[33] Taken together, these 3 secondary structures made with graphite epoxy materials weighed 1,500 pounds and represented a 450-pound weight reduction over the aluminum components.[34]

Three other medium primary structures were designed for the ACEE program: the vertical fin, the horizontal stabilizer, and vertical stabilizer. The "medium primary" classification meant that other components were attached to them (and they provided the aircraft with stability), so they were more critical to a safe flight than the secondary structures were. The vertical fin is at the rear of the airplane, where it contributes aerodynamic directional stability.

Design of a composite vertical fin for the Lockheed L-1011 started in 1975, and the project was then transferred to the ACEE program once

33. "Composite Ailerons Readied for L-1011s," *Aviation Week & Space Technology*, Oct. 13, 1980, p. 27.

34. "ACEE Program Overview," July 31, 1979, Box 239, Division 8000, NASA Glenn archives.

underway in 1976.[35] The development was plagued by several problems when the composite materials failed prior to reaching ultimate load, and as a result, it never progressed beyond the static testing stage under ACEE. Though it never took flight, this was a 7-foot by 23-foot structure and, at 780 pounds, represented a 22.6-percent weight savings. The next medium primary structure designed was the horizontal stabilizer. This is a fixed surface at the rear of the airplane that provides longitudinal stability.[36] Designed for a Boeing 737, it too experienced structural failures during ground tests, but these were corrected, and the FAA certified the component in August 1982. On April 11, 1984, the first composite primary structure went into service, representing 28.4-percent weight savings.[37] The final medium primary structure was the vertical stabilizer. Located at the back of the airplane, it is used to control yaw, or the rotation of the vertical axis.[38] Designed for the Douglas DC-10, the vertical stabilizer provided a 22.1-percent reduction in weight, but it too experienced several production problems and failed a ground test. After the failure, engineers incorporated a different structure, and though it took much more time to develop than expected, the FAA certified it in 1986, and commercial flight commenced in January 1987.[39]

Langley engineers wrote a number of computer programs to aid in the design and analysis of these composites. PASCO analyzed composite panels and helped determine their material strength. VIPASA provided

---

35. T. Alva, J. Henkel, R. Johnson, B. Carll, A. Jackson, B. Mosesian, R. Brozovic, R. O'Brien, and R. Eudaily, *Advanced Manufacturing Development of a Composite Empennage Component for L-1011 Aircraft* (Washington, DC: NASA CR-165885, May 1982).
36. "Boeing 737 with New Stabilizer Makes First Flight," *Aviation Week & Space Technology,* Oct. 13, 1980, p. 37.
37. R.B. Aniversario, S.T. Harvey, J.E. McCarty, J.T. Parsons, D.C. Peterson, L.D. Pritchett, D.R. Wilson, and E.R. Wogulis, *Design, Ancillary Testing, Analysis and Fabrication Data for the Advanced Composite Stabilizer for Boeing 737 Aircraft,* vol. 2 (Washington, DC: NASA CR-166011, Dec. 1982).
38. C.O. Stephens, *Advanced Composite Vertical Stabilizer for DC-10 Transport Aircraft* (Washington, DC: NASA CR-173985, July 1978).
39. "New Tests of DC-10 Composite Stabilizer Set," *Aviation Week & Space Technology,* May 6, 1985, p. 94. Dow, "The ACEE Program and Basic Composites Research at Langley Research Center (1975 to 1986)" (Washington, DC: NASA RP-1177, 1987), p. 1. L.F. Vosteen, *Composite Structures for Commercial Transport Aircraft* (Washington, DC: NASA TM-78730, June 1978).

data on buckling and vibration and worked in conjunction with PASCO. CONMIN was a nonlinear mathematical programming technique that assisted in sizing issues.[40] Three years later, another program, POSTOP, assisted in the design of composite panels by analyzing compression, shear, and pressure on the materials.[41] Temperature effects were also included. Other design and analysis studies used traditional mathematics and experimentation. Extensive failure studies were undertaken to help ensure the durability of the composite structures. One type of study analyzed what happened when surfaces cracked and how that compromised the safety of the airplane.[42] Resulting experiments looked at repair techniques for these composite structures when cracks and other tears appeared.[43]

Engineers also designed long-term environmental studies to determine the possible effects of environmental exposure on the composites. One concern was that the composites would degrade over time because of ultraviolet light. Another concern was whether they would absorb moisture. Tests included composite panels placed on airport rooftops at Langley and in San Diego, Seattle, São Paulo, and Frankfurt. These took into account geographical location, solar heating effects, ultraviolet degradation, and test temperatures.[44] Other studies evaluated components in flight. Richard A. Pride, who headed the program at Langley, found that after 3 years, "No significant degradation has been observed in residual strength."[45] Longer-term studies, up to 10 years, indicated that composites did not degrade over time given normal use and environmental exposure.[46]

---

40. M.S. Anderson, W.J. Stroud, B.J. Durling, and K.W. Hennessy, *PASCO: Structural Panel Analysis and Sizing Code* (Washington, DC: NASA TM-80182, Nov. 1982).
41. S.B. Biggers and J.N. Dickson, *POSTOP: Postbuckled Open-Stiffener Optimum Panels* (Washington, DC: NASA CR-172260, Jan. 1984).
42. C.D. Babcock and A.M. Waas, *Effect of Stress Concentrations in Composite Structures* (Washington, DC: NASA CR-176410, Nov. 1985).
43. J.W. Deaton, *A Repair Technology Program at NASA on Composite Materials* (Washington, DC: NASA TM-84505, Aug. 1982).
44. D.J. Hoffman, *Environmental Exposure Effects on Composite Materials for Commercial Aircraft* (Washington, DC: NASA CR-165649, Aug. 1980).
45. R.A. Pride, "Interim Results of Long Term Environmental Exposures of Advanced Composites for Aircraft Applications," *Proceedings of the 11th Congress of the International Council of the Aeronautical Sciences,* vol. 1, Sept. 1978, pp. 234–241.
46. R.L. Coggeshall, *Environmental Exposure Effects on Composite Materials for Commercial Aircraft* (Washington, DC: NASA CR-177929, Nov. 1985).

Despite the success of these studies, there was one important environmental concern that threatened to halt the composites program and for a time did ground all composite flight-testing. Because carbon fibers were a main component of these composites, flight over population centers was an environmental issue. The risk was to everyday electrical systems that could potentially be damaged through exposure to the accidental release of carbon fibers into the air through an accident or crash. There was a possibility that fibers released from composite aircraft materials could interfere with electrical systems on the ground (because the fibers can conduct electricity), causing them to fail. The concern spanned from the mundane—a toaster or television—to the critical—air traffic control equipment or nuclear powerplants. The fibers were so light that they could be easily blown and distributed in the air by an explosion, affecting a wide area. Moderate winds could spread them tens of miles. The airline industry was concerned because it would then be liable for replacing all the failed electronics equipment.[47] A major ACEE investigation, the Carbon Fiber Risk Assessment, was launched to determine the significance of this threat.[48] It was headed at Langley by Robert J. Huston, the Program Manager of the Graphite Fibers Risk Analysis Program Office.

After extensive research at Langley, engineers concluded that the threat would be negligible.[49] For example, 0.00339 televisions out of 100 would fail. Only 0.00171 toasters would be affected out of 100. For more critical equipment, the predictions were also low, only 0.005 out of 100 types of air traffic control equipment, or 0.016 out of 100 ground computer installations. After more than 50 technical reports, NASA predicted that carbon fiber accidents would only cause $1,000 worth of damage in 1993, and the absolute worst-case scenario would be a $178,000 loss occurring every 34,000 exposures.[50] Compared with other possible air transportation threats, the carbon fiber risk was simply nonexistent.

While the ACEE composites program lasted 10 years, from 1976 to 1985, it ended before achieving its major goal of developing wings and

---

47. "Composites Programs Pushed by NASA," *Aviation Week & Space Technology,* Nov. 12, 1979, p. 203.
48. "Carbon Fiber Hazard Concerns NASA," *Aviation Week & Space Technology,* Mar. 5, 1979, p. 47.
49. Risk Analysis Program Office, *Risk to the Public From Carbon Fibers Released in Civil Aircraft Accidents* (Washington, DC: NASA SP-448, 1980).
50. Ibid.

fuselages with composite materials, the stated goal of the program, because the wing and fuselage represented 75 percent of the weight of the airplane. Wings and fuselages made of composites would have achieved significant weight savings and fuel economy.[51] There were several reasons that these were never developed by ACEE. First was the amount of time and resources devoted to the Carbon Fiber Risk Assessment. Unanticipated at the start of the project, this potential problem became a serious threat to the use of composites. Therefore, it was necessary to prove that there was little risk in their use. After this setback, NASA was finally able to devote all of its attention to wings and fuselages in 1981, but engineers took a different approach to their development than they did with the previous components. Whereas before NASA had developed composites that replaced entire metal components on aircraft, it now decided to try to incorporate composite pieces into the fuselage (a section barrel) and wing (short-span wing box). Boeing studied the damage tolerance of composite wings, the threat posed by lightning strikes, and an evaluation of their fuel sealing capabilities.[52] Lockheed examined acoustic issues, such as how noise was transmitted through flat, angular, composite panels and how to reduce it.[53] By this time, the ACEE program and its funding were nearly at its end, so the ultimate goal of composite wings and fuselages was never attained.[54]

Nevertheless, the success ACEE had with secondary components was called "almost revolutionary." One observer said this 10-year period represented the "golden age of composites research in the United States."[55] ACEE became a "strategic center of gravity" in this golden age, and its achievements in secondary structures were vitally important in

---

51. "Composite Wing," *Aviation Week & Space Technology,* Dec. 28, 1981, p. 12.
52. C.F. Griffin, *Fuel Containment and Damage Tolerance in Large Composite Primary Aircraft Structures* (Washington, DC: NASA CR-166083, Mar. 1983).
53. J. Lameris, S. Stevenson, and B. Streeter, *Study of Noise Reduction Characteristics of Composite Fiber-Reinforced Panels* (Washington, DC: NASA CR-168745, Mar. 1982).
54. P.J. Smith, L.W. Thomson, and R.D. Wilson, *Development of Pressure Containment and Damage Tolerance Technology for Composite Fuselage Structures in Large Transport Aircraft* (Washington, DC: NASA CR-178246, 1986). A.C. Jackson, F.J. Balena, W.L. LaBarge, G. Pei, W.A. Pitman, and G. Wittlin, *Transport Composite Fuselage Technology—Impact Dynamics and Acoustic Transmission,* (Washington, DC: NASA CR-4035, 1986).
55. Dow, "The ACEE Program and Basic Composites Research at Langley Research Center (1975 to 1986)," NASA RP-1177, 1987, p. 5.

introducing a new type of material as an alternative to the traditional metal and aluminum used in airplanes. The Composite Primary Aircraft Structures program had several very significant results over its lifespan.[56] It produced 600 technical reports and provided a cost estimate for developing these materials and a confidence in their durability and long-term use. Composites received certification by the FAA, as well as general acceptance by the airline industry. Overall, it's estimated that the ACEE program was responsible for accelerating the use of composites in the airline industry by at least 5 to 10 years. Langley continued to track the composites it developed even after the ACEE program concluded, and 350 composites reached 5.3-million flight-hours in 1991 and were still operational.

According to Herman Rediess, one of the initial ACEE task force members, "Many of things that we were talking about at the time are now just so standard that people hardly even remember that they came out of ACEE." Prior to the ACEE program, aircraft manufacturers were reluctant to investigate the opportunities these composites offered because of costs and unknown performance capabilities. But, as Rediess now reflects, "It's a major, major aspect of our commercial transports. It has really paid off in terms of weight savings, and in that weight is fuel."[57] By the 1990s, these composite materials resulted in a fuel efficiency savings of 15 percent.[58] As one observer concluded at a 1990 conference on composite materials, "The NASA Aircraft Efficiency Program provided aircraft manufacturers, the FAA, and the airlines with the experience and confidence needed for extensive use of composites in . . . future aircraft."[59]

Since the end of the ACEE program, the use of composites has increased, though not as dramatically as first imagined. While the weight savings and fuel efficiency were undeniable, their mass implementation was offset by the cost of producing them, compared with metal and

---

56. Richard G. O'Lone, "Industry Tackles Composites Challenge," *Aviation Week & Space Technology,* Sept. 15, 1980, p. 80.

57. Interview with Herman Rediess by Bowles, June 6, 2008.

58. Mark A. Chambers, *From Research to Relevance: Significant Achievements in Aeronautical Research at Langley Research Center (1917–2002)* (Washington, DC: NASA NP-2003-01-28-LaRC, 2003), p. 15.

59. R.C. Madan and M.J. Shuart, "Impact Damage and Residual Strength Analysis of Composite Panels with Bonded Stiffeners," *Composite Materials: Testing and Design,* vol. 9, S.P. Garbo, ed. (Philadelphia: American Society for Testing and Materials, 1990), p. 64.

This X-29 research aircraft in flight over California's Mojave Desert shows its striking forward-swept wing and canard design. The X-29 demonstrated the use of advanced composites in aircraft construction. Two X-29 aircraft flew at the Ames-Dryden Flight Research Facility from 1984 to 1992. (NASA Dryden Flight Research Center Photo Collection.)

aluminum structures. They are also more expensive to certify for flight readiness.[60] As fuel costs increase in the 21st century, however, the economic returns for lighter aircraft will become more valuable, and composites will take on greater significance. Today, the military has surpassed commercial aviation in the use of composites. For example, composites account for 38 percent of the weight of an F-22 but only 10 percent of a Boeing 777, which has the highest composite percentage of any commercial aircraft.[61]

The new Boeing 787 Dreamliner may become the first major commercial aircraft with composites comprising the majority of its materials, as the company is planning for 50 percent of primary structures, including

---

60. Alan Baker, Stuart Dutton, and Donald Kelly, eds., *Composite Materials for Aircraft Structures*, 2nd ed. (Reston, VA: American Institute of Aeronautics and Astronautics, 2004), p. 1.
61. Chambers, *Concept to Reality*.

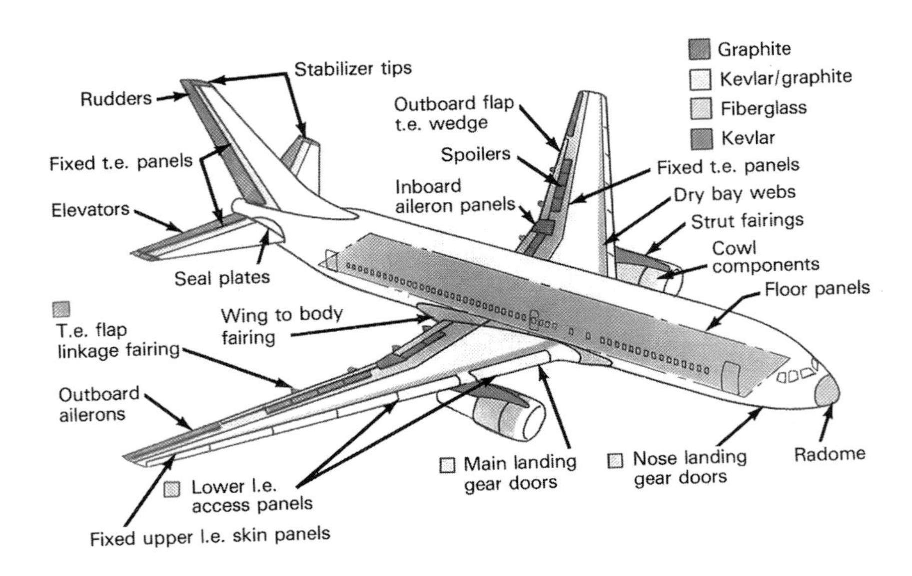

The above graphic demonstrates the composite components of the Boeing 767. Courtesy of Joseph Chambers.

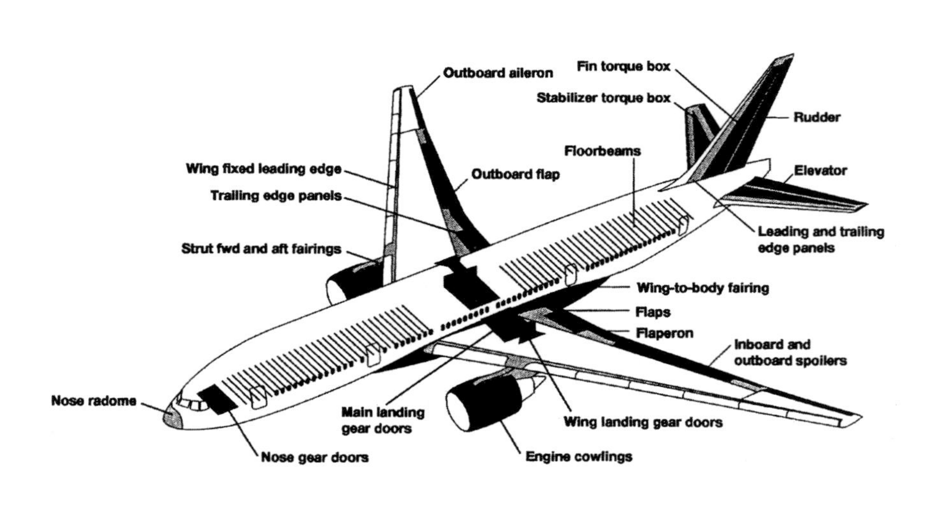

The above graphic demonstrates the composite components of the Boeing 777. Courtesy of Joseph Chambers.

Composite aircraft Lancair Columbia and Cirrus SR20. Courtesy of Joseph Chambers.

fuselage and wing, to be composites.[62] The general aviation community has also benefited from composites. For example, small personal-owner aircraft and homebuilt aircraft, with designer Burt Rutan taking the lead, have taken advantage of composites technology. Business-class aircraft such as Beech Aircraft (now Raytheon Aircraft Company) has developed an all-composite aircraft known as the Lancair Columbia 300 and the Cirrus SR20.

The ACEE composites program was a success because, according to Jeffrey Ethell, it "demolished the fear factor surrounding the new materials, which have entered the real world of transport aviation."[63] ACEE served as an encouraging point of departure for industry entering the world of composites. The program took materials that were untested, unusual, and exotic, and it transformed them into certified and usable structures on commercial and military aircraft. According to Joseph Chambers, "The legacy of the ACEE Program and its significant contributions to the

---

62. See "Boeing Dreamliner Will Provide New Solutions for Airlines, Passengers," as found at *http://www.boeing.com/commercial/787family/background.html,* accessed Sept. 2, 2009.
63. Ethell, *Fuel Economy in Aviation,* p. 74.

acceleration, acceptance, and application of advanced composites has become a well-known example of the value of Langley contributions to civil aviation. In the best tradition of NASA and industry cooperation and mutual interest, fundamental technology concepts were conceived, matured, and efficiently transferred to industry in a timely and professional manner."[64]

Advanced Aerodynamics—Energy Efficient Transport

Another important ACEE airframe technology program was one Langley engineers Robert W. Leonard and Richard D. Wagner called a "somewhat arbitrarily termed 'Energy Efficient Transport.'"[65] Like the composites program, this ACEE project was also to be managed by Langley, and it promised to be of great importance to industry. Unlike the composites program, whose objective was to focus on a single technology development that promised significant fuel savings, the Energy Efficient Transport project planned to achieve fuel efficiency through a number of aerodynamics advances. These included the following areas of research: supercritical wings, winglets, nacelle aerodynamic and inertial loads, wing and tail surface coatings, laminar flow control, and active controls. NASA's Langley Research Center partnered with Boeing Commercial Airplane, Douglas Aircraft, and Lockheed-California to analyze, design, test, and assess these advanced aerodynamic concepts.[66]

Most commercial airplanes fly at what is known as transonic speeds. This is an aeronautics term for velocities just below and above the speed of sound (Mach 0.8 to 1.2). "The transonic regime," as Roger Bilstein said, "had beguiled aerodynamicists for years." Despite being the speed of choice for modern aircraft, those cruise speeds present numerous aerodynamic challenges.[67] At these speeds, both subsonic (less than the speed

---

64. Chambers, *Concept to Reality*.

65. Leonard and Wagner, "Airframe Technology for Energy Efficient Transport Aircraft," *Aerospace Engineering and Manufacturing Meeting of the Society of Automotive Engineers, Nov. 29–Dec. 2, 1976*.

66. David B. Middleton, Dennis W. Bartlett, and Ray V. Hood, *Energy Efficient Transport Technology* (Washington, DC: NASA RP-1135, Sept. 1985). This is an excellent concluding summary to the EET program and includes a comprehensive bibliography.

67. Bilstein, *Testing Aircraft, Exploring Space*, pp. 106–107.

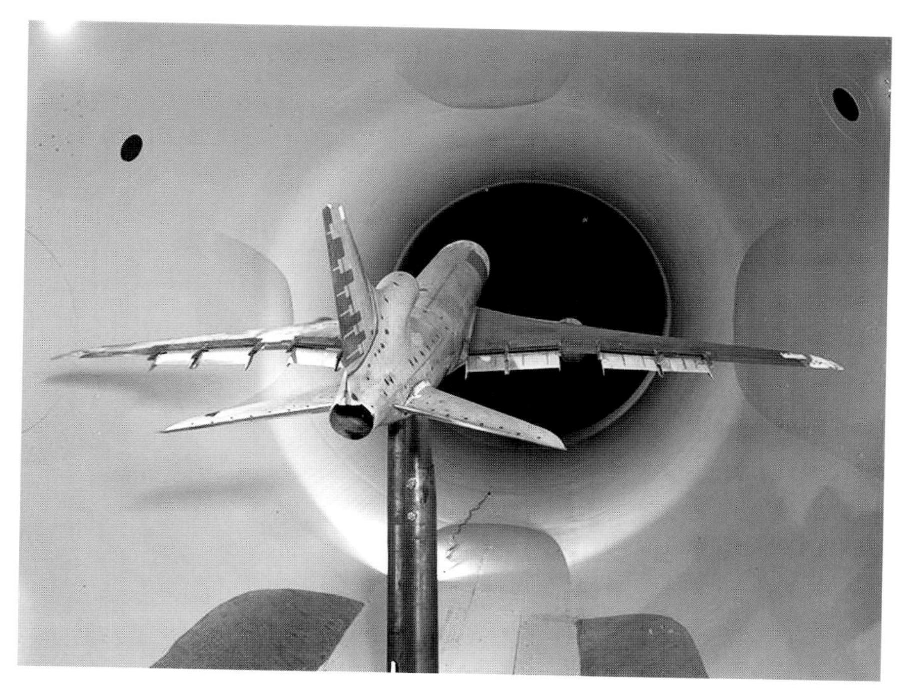

Lockheed L-1011 EET model used in testing of the Energy Efficient Transport project at Langley Research Center (January 5, 1982). (NASA Ames Research Center [NASA ARC].)

of sound) and supersonic (more than the speed of sound) airflow patterns exist over the aircraft simultaneously, so even if an airplane is flying at subsonic speeds, airflow over certain sections of the wing might reach supersonic levels, forming strong shock waves on the upper surfaces of the wings and resulting in a dramatic increase in drag. This problem is known as the "sound barrier." As one observer said, "The barrier was conquered [in 1947] with brute force, but the trick now is to subdue it quietly and efficiently."[68] Solving the challenge, known as the "supercritical" Mach number, was an important problem.[69] Engineers knew that if they could solve it, they would significantly improve cruise performance and increase fuel efficiency. It was to this task that Langley engineer Richard Whitcomb first applied himself in the 1960s. After several years of

68. Ethell, *Fuel Economy in Aviation,* p. 78.

69. Bill Siuru and John D. Busick, *Future Flight: The Next Generation of Aircraft Technology,* 2nd ed. (Blue Ridge Summit, PA: TAB/AERO, 1994), p. 36.

Richard Whitcomb looks over a model of the Chance Vought F-8 aircraft incorporating his supercritical wing (July 1, 1970). (NASA Langley Research Center [NASA LaRC].)

research and extensive wind tunnel studies, he redesigned the wing shape with a flatter upper surface, which reduced the strength of shock waves. A downward sloping curve at the wing's trailing edge increased the lift. Because the supercritical wing could be thicker than a conventional wing, the aspect ratio of the wing could be increased to reduce the drag, and the wing sweep could be decreased for more efficient cruise. The "supercritical wing" was born. In 1972, after 12 test flights, Whitcomb said, "I feel confident we've reached a milestone in the program."[70]

To take advantage of Whitcomb's work, NASA needed an incentive to perform further flight tests and incorporate it into a commercial transport that would be both aerodynamically and structurally sound. This incentive came with the fuel crisis, and the supercritical wing became part of the EET project. Langley engineers began generating a database of wing variables in wind tunnels that tested the various effects of thickness, camber

70. Richard Whitcomb, quoted in Richard Witkin, "NASA Says the Supercritical Wing is Passing Tests," *New York Times*, Mar. 5, 1972.

NASA selected a Vought F-8A Crusader as the testbed for an experimental supercritical wing. (January 1, 1972). (NASA Dryden Flight Research Center [NASA DFRC] Photo Collection.)

(the wing's curvature), sweep, and aspect ratio (a measure of the wing's ratio of span to area). The results of their research led to the adoption of this wing in a variety of aircraft. Industry followed Whitcomb's lead with its own supercritical wing designs. Boeing incorporated a version of the wing in the Boeing 767 in 1981 and the Boeing 777 in 1995. James Hansen has called the 777's wing the "most aerodynamically efficient airfoil ever developed for subsonic commercial aviation."[71] The wing's success can be traced directly back to the pioneering work performed by Whitcomb and the ACEE engineers.

The supercritical wing was not Whitcomb's only inspiration. In 1974, he developed a new idea, known as winglets. While the supercritical wing promised fuel efficiency in the future when new aircraft were built, the winglets were important because they could be immediately retrofitted.

Looking and acting like a vertical sail, they took advantage of the swirling vortex of airflow around the tip of the wing. Whitcomb published

---

71. Hansen, *The Bird is On the Wing,* p. 196.

the results of his study in July 1976 and promised a 4- to 8-percent drag reduction. He confidently predicted, "Just as sure as the sun rises, the next new commercial transport aircraft will have winglets."[72] Since the whirlpool of air around the wingtip was different for every airplane, it was left to the aircraft manufactures to design and test specific winglets for their planes. To encourage their adoption, NASA and the EET program cosponsored industry flight tests on aircraft. The first, between 1978 and 1979, included research with Douglas Aircraft on its DC-10. The success in reducing fuel consumption was quickly apparent; Robert Leonard, Langley's ACEE Project Manager, said, "Frankly, the winglet looks very promising on the DC-10."[73] Douglas designers incorporated winglets into their new MD-11 development in 1986.[74] Very quickly, the entire industry realized the importance of the winglet.[75]

Other issues caused by structural reinforcement for flutter and gust loads required solutions.[76] One approach to providing structural weight reduction while maintaining safety margins was a computer-assisted advance called Active Controls Technology (ACT), also known as a Control-Configured Vehicle (CCV). While ACT technology had been investigated prior to ACEE, this program served to increase dramatically confidence and industry acceptance. The ACT system used an onboard computer system to control surfaces on the trailing edges of the wings and on the tail sections of the aircraft. The computer acted independently from the pilot, working to minimize the aircraft's structural loads when it encountered turbulence or making a tight turn while maintaining a sufficient flutter margin. To achieve this, sensors on the surfaces of the aircraft sent feedback to the computer, which in turn could send compensating signals to the control surface actuators. Computers, not pilots, were best suited to handle these controls because turbulence is a random, time-

---

72. Whitcomb, quoted in Warren C. Wetmore, "Langley Presses Fuel Efficiency Programs," *Aviation Week & Space Technology,* Nov. 10, 1975, p. 68.

73. Jerry Mayfield, "Energy Efficiency Research Growing," *Aviation Week & Space Technology,* Nov. 12, 1979, p. 119.

74. James J. Haggerty, "Winglets for the Airlines," *Spinoff 1994,* pp. 90–91.

75. Hansen, *The Bird Is on the Wing,* p. 200.

76. R.V. Hood, Jr., "The Aircraft Energy Efficiency Active Controls Technology Program," *Guidance and Control Conference, Aug. 8–10, 1977, Hollywood, FL,* AIAA Paper 77-1076, p. 279.

A KC-135 with winglets in flight over the San Gabriel Mountains, south of Edwards. (January 1, 1979). (NASA Dryden Flight Research Center [NASA DFRC] Photo Collection.)

dependent phenomenon, and the electronic system can react much faster than a human pilot can. The sensors and the computer were able to communicate to rudders, elevators, and ailerons within a split second to adjust correctly for these disturbances.[77] Hindering the development of such a system were a lack of confidence that the design was possible and a belief that they were not cost effective. Langley engineers worked with counterparts at Douglas, Lockheed, and Boeing to solve these problems and install active controls on several types of specific airplanes. The results of these studies proved that an ACT airplane required an investment of $600,000, with a 25-percent return on investment (based on fuel prices in the early 1980s). The FAA also concluded that they were flightworthy, and that no single failure in the system would result in the loss of control of the aircraft. Pan American World Airways purchased the first aircraft with active controls (L-1011-500) and then began to retrofit active controls to all planes of this type in its fleet.[78]

Another important aerodynamics advance explored under the EET program focused on airframe/propulsion integration. The main effort in

---

77. Jeffrey M. Lenorovitz, "L-1011 Active Control System Tested," *Aviation Week & Space Technology,* Sept. 19, 1977, p. 26.
78. Ethell, *Fuel Economy in Aviation,* p. 96.

this area was the Nacelle Aerodynamic and Inertial Loads (NAIL) program directed jointly by Langley and Lewis Research Center. Engineers knew that the most critical period of deterioration for aircraft engine efficiency occurs during the initial period of its life. After the engine reaches approximately 1,000 flights, this deterioration levels off substantially. The goal of the NAIL program was to provide as much data as possible on the early life of a jet engine to determine the causes for the decreases in efficiency. The Centers partnered with Boeing and Pratt & Whitney, and a NAIL engine was constructed, flown, and then disassembled and inspected. The test flights revealed that the highest "flight loads," or wear, occurred at low speeds, high angles of attack, and high engine airflow, conditions most typically occurring at takeoff.[79] The conclusions served as the basis for future nacelle redesigns that would have a greater ability to withstand flight-load wear and tear, specifically during these periods of flight.

There were two areas of the EET program that overlapped with other ongoing ACEE investigations at Langley. Much like the composites program, the Aircraft Surface Coatings program explored the use of new materials that would improve the surface smoothness of aircraft. The Apollo spacecraft had used Kapton, a film polyimide, as a coating, which reduced drag, decreased maintenance, and offered increased protection. Similar advantages were sought for aircraft surfaces. Langley engineers identified elastomeric polyurethane coatings such as CAAPCO and Chemglaze and tested them on a Continental Airlines Boeing 727 used by Air Micronesia. Micronesia was selected because of its high rainfall environment, which typically degrades surface coatings. The engineers found that these materials produced a small decrease in drag and at the same time increased protection from corrosion.[80] One question the EET program left unanswered was whether the polyurethane would work equally well to reduce drag on larger winglike surfaces with curvatures.

A second EET program with similarities to another ACEE program was laminar flow (see chapter 5 for a complete description of laminar flow). EET engineers performed natural laminar flow studies that resulted in some

---

79. *Nacelle Aerodynamic and Inertial Loads (NAIL) Project-Summary Report* (Washington, DC: NASA CR-3585, 1982).
80. *Aircraft Surface Coatings—Summary Report* (Washington, DC: NASA CR-3661, 1983).

successes. When analyzed, a 757 achieved a significant natural laminar flow, improving fuel efficiency on a Mach 0.8 flight over 2,400 miles.[81]

The EET programs—supercritical wings, winglets, nacelle aerodynamic and inertial loads, wing and tail surface coatings, laminar flow control, and active controls—were successful in reaching the goals of the ACEE program. EET was the focus of nearly 150 technical reports, which serve as a comprehensive database describing the new ideas that were evaluated and proved viable. These reports expressed an overall confidence that EET would result in the production of new airplanes that would attain at least 15 to 20 percent more fuel efficiency than those currently in production.[82] Of these, James Kramer, who initially headed the ACEE Committee, said "the major visible EET results" of this program were the winglets, supercritical wings, and the active control technologies. The advanced aerodynamics investigations of ACEE were a success.[83]

Ironically, some of these fuel-saving technologies diffused more quickly among European nations. In the early 1980s, Richard Wagner, a Langley ACEE manager, said he was flying a French-made Airbus A310 to Israel and, to his great surprise, when he looked out his window while the plane was still on the tarmac, on the tip of the wing he saw a winglet. It was actually the Israelis who were the first to apply winglets on the Westwind. Although they had been in use on smaller business jets in the United States, this was the first time Wagner had seen winglets on a commercial transport, where the winglets had their greatest advantage. Wagner concluded with some remorse, "So it seems like the Europeans, in my own personal observation, may have capitalized more upon the ACEE program results than our own American companies."[84]

A further concern that Langley managers articulated at the start of the ACEE program was the threat to American dominance of aircraft manufacturing on the world stage. By 1982, Eastern Air Lines had purchased 34 Airbus A300 transports. This moved Airbus into second place

---

81. *Natural Laminar Flow Airfoil Analysis and Trade Studies—Final Report* (Washington, DC: NASA CR-159029, 1979).

82. Middleton, Bartlett, and Hood, *Energy Efficient Transport Technology,* p. 16.

83. James J. Kramer, "Aeronautical Component Research," *Advisory Group for Aerospace Research and Development* (Report No. 782, 1990), p. 5-4.

84. Interview with Wagner by Bowles, June 30, 2008.

internationally in terms of aircraft manufacturing, putting it ahead of the American Douglas and Lockheed companies. The international challenge was growing, but because the Energy Efficient Transport and the composites program were already showing important fuel-efficiency returns, many believed the United States would remain competitive despite the growing challenge from Airbus and European governmental support. Robert Leonard, of Langley, believed one major reason was that "fuel efficiency will continue to dominate purchase decisions by the world's airlines."[85] Assisting in this effort were the NASA engineers at Lewis Research Center in Cleveland, OH, who focused on innovative propulsion project to further improve fuel efficiency.

85. Leonard, "Fuel Efficiency Through New Airframe Technology," NASA TM-84548, Aug. 1982.

Chapter 3:

Old and New Engines at Lewis

Pan American World Airways was for a time the most influential airline organization and known quite simply as the airline that "taught the world to fly."[1] Begun in the late 1920s through a legendary partnership between Juan Trippe and Charles Lindbergh, it sought to promote international commercial air transport in the United States. It was widely successful and through the early 1970s led the world in the commercial transport industry. Because of the increasing price of oil, however, the United State's largest airline suffered a major setback in 1973, and the company was driven to the edge of bankruptcy in 1974.[2] L.H. Allen, its vice president and chief engineer, said that when "fuel prices for Pan Am . . . reached a staggering 40 cents per gallon," fuel efficiency then ranked as the "single most important factor in aircraft operations today."[3] Allen tried to offset these devastating forces by working with NASA's Lewis Research Center and its Aircraft Energy Efficiency (ACEE) program to devise new ways to make aircraft engines more efficient.

Like the Langley Research Center, NASA's Lewis Research Center in Cleveland, OH, operated its own specialty fuel-efficiency research programs under ACEE. The first of these propulsion projects, Engine Component Improvement (ECI)—with Pan Am serving as one of its chief

---

1. Asra Q. Nomani, "Pan Am Seeks Chapter 11 Shield, Fuel Costs are Cited," *Wall Street Journal,* Jan. 9, 1991, p. A3.
2. Jack Egan, "Pan Am Warns Survival in Fuel Crisis May Depend on U.S. Aid," *Los Angeles Times,* Dec. 5, 1973, p. B10. "Pan Am Loss Jumps More Than 4 Times to $81 Million," *Los Angeles Times,* Feb. 6, 1975, p. E14.
3. N.B. Andersen and L.H. Allen, Engine Component Improvement Program, Airline Evaluation, Dec. 19, 1980, Report of Meeting No. 10, Box 208, Division 8000, NASA Glenn archives.

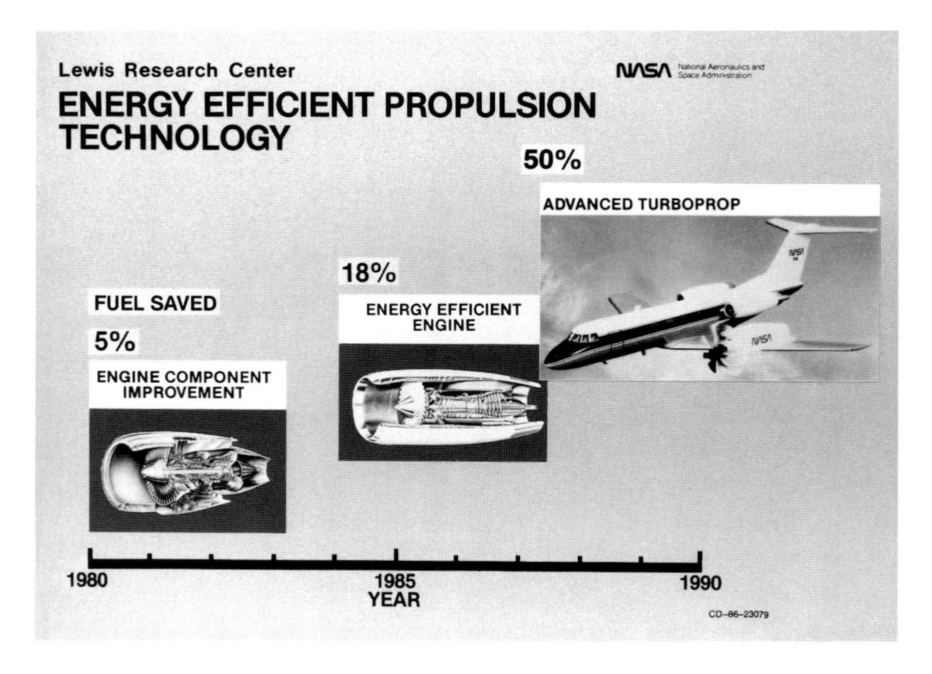

The three ACEE projects at Lewis Research Center. (NASA Glenn Research Center [NASA GRC].)

independent evaluators and main program supporter—focused on improving existing turbofan engines through the redesign of selected engine components that were most prone to wear.[4] An attempt to "cure sick engines," it was the least technically challenging of the three Lewis ACEE projects and aimed for a 5-percent improvement in fuel efficiency. The second project, the Energy Efficient Engine (E^3), involved building "a brand new engine from scratch" and offered a far greater payoff—a 10- to 15-percent increase in fuel efficiency.[5] These two engine projects became Lewis's most significant contribution to improving fuel efficiency for the Nation's commercial aircraft. (A third Lewis project, the Advanced Turboprop, the most controversial of all, will be addressed in a later chapter.)

While Pan Am's collaboration on ACEE was successful, it was not enough to save the company, which declared bankruptcy in January 1991.

4. Donald L. Nored to director of aeronautics, Feb. 8, 1979, Box 244, Division 8000, NASA Glenn archives.

5. Interview with Nored by Dawson and Bowles, Aug. 15, 1995.

Langley engineers at the Structural Research Laboratory designing the NACA's new engine lab in Cleveland. Among those pictured are: Addison Rothrock, George Darchuck, Harold Friedman (at the front and center with his back to the camera), and Nick Nahigyan (across table from Friedman) (April 21, 1941). (NASA Glenn Research Center [NASA GRC].)

The reason most often cited for its demise, according to historian William Leary, is that it was never able to recover from the "the world oil crisis."[6]

FROM ENGINES TO ENERGY—LEWIS RESEARCH CENTER

Established during World War II as an aircraft engine research laboratory, Lewis became the third laboratory of the National Advisory Committee for Aeronautics, following Langley and Ames.

---

6. William M. Leary, "Sharing a Vision: Juan Trippe, Charles Lindbergh, and the Development of International Air Transport," *Realizing the Dream of Flight: Biographical Essays in Honor of the Centennial of Flight, 1903–2003,* Dawson and Bowles, eds. (Washington, DC: NASA SP-4112, 2005), p. 76. Likewise, Robert Gandt, a pilot who flew with Pan Am for 26 years, said the company's failure was due to the "complex life and times" caused by the energy crisis of the 1970s, rather than a failure of leadership. Robert Gandt, *Skygods: The Fall of Pan Am* (New York: William Morrow and Company, 1995).

Bruce Lundin in 1960 at the Rocket Laboratory. Lundin investigated heat transfer and worked to improve the performance of World War II aircraft engines. From 1969 to 1977, he was Director of the Lewis Research Center. (NASA Glenn Research Center [NASA GRC].)

Lewis engineers pursued aircraft engine research in the national interest—often over the objections of the engine companies, who perceived the Government as interfering with the normal forces of supply and demand. During the early years of the Cold War, the laboratory participated in engine research and testing to assist the engine companies in developing the turbojet engine. After the launch of Sputnik, the laboratory focused on a new national priority—rocket propulsion research and development. All work on air-breathing engines ceased for nearly 10 years. The return to aircraft engine research coincided with drastic reductions in staff, mandated by cuts in NASA's large-scale space programs. The mass exodus of nearly 800 personnel in 1972 sparked an effort to redefine the Center's mission and find new sources of funding. One year later, the Nixon Administration reduced the NASA budget by $200 million.[7] This coincided with OPEC's oil embargo and galvanized the Center's Director, Bruce Lundin, to look for ways to use its propulsion expertise to help solve the energy crisis.[8]

---

7. Harold M. Schmeck, Jr., "NASA Cuts Programs to Save $200-Million in Current Budget," *New York Times,* Jan. 6, 1973.
8. Dawson, *Engines and Innovation,* p. 204.

Lewis Research Center in 1968. (NASA Glenn Research Center [NASA GRC].)

Lewis engineers were exploring a variety of alternative energy programs, and Virginia Dawson characterized its new focus as "Lewis turns earthward." These efforts began in the early 1970s with the NASA Volunteer Air Conservation Committee, headed by Louis Rosenblum and J. Stuart Fordyce. They were inspired by the tragic symbol of a polluted Cleveland, which became a national joke after the literal burning of its Cuyahoga River. Then Robert Hibbard began a graduate seminar with students from area universities, which focused on ways to develop cleaner engines and other environmental issues.[9] In 1971, Lewis established its Environmental Research Office, set up monitoring stations throughout the city, and worked with the Environmental Protection Agency (EPA) to study water pollution in Lake Erie. In the early 1970s, Lewis engineers also initiated research using its nuclear test reactor at Plum Brook Station, irradiating over 1,000 samples per year for the EPA.[10] According to Dawson, emerging from these programs "were the seeds from which an

---

9. "Pollution Course Taught at Lewis," *Lewis News,* Dec. 31, 1970.

10. Bowles, *Science in Flux: NASA's Nuclear Program at Plum Brook Station, 1955–2005* (Washington, DC: NASA SP-4317, 2006), p. 144.

Windmill project conducted by Lewis Research Center (September 3, 1975). (NASA Glenn Research Center [NASA GRC].)

entirely new effort would grow."[11] These efforts were soon followed by investigations into alternative energy sources—wind, solar, and electric.

In 1974, Lewis received $1.5 million for a wind-energy program from the National Science Foundation and the Energy Research and Development Administration (ERDA). As a result, the Lewis-managed Plum Brook Station eventually built experimental windmills for research. With 2 massive 62-foot propeller blades, the first 125-foot windmill was capable of generating 100 kilowatts. At the time, it was the second largest

11. Dawson, *Engines and Innovation,* p. 204.

windmill ever constructed in the United States. One engineer who worked on the Lewis windmills predicted that the country would soon see "hundreds of thousands of windmills generating electricity across the United States."[12] The most impressive of those built by Lewis engineers was a commercial wind turbine generator in Hawaii in 1988, which was then the world's largest.[13]

A program in solar cell technology development followed on the windmill project's heels, along with increasing funding for various energy-related programs by ERDA and its successor, the Department of Energy.[14] Though Lewis lost out on a bid for a $35-million Federal solar research institute, its growing expertise in alternative energy was becoming well known. In 1978, Lewis's engineers were consulted in building the world's first solar-power system for a community—the 96 residents of the Papago Indian village about 100 miles northwest of Tucson. Louis Rosenblum designed the solar array and helped install it. His system replaced the Papago tribe's kerosene lighting in 16 homes, a church, and a tribal feast house.[15]

The creation of an electric automobile was another Lewis project. Known as the Hybrid Vehicle Project, its engineers researched several experimental concepts to achieve increased fuel efficiency and decreased emissions and address a growing national need caused by its energy dependence.[16] These primarily electric vehicles were charged by an outside source. Lewis engineers completed their initial plan in 1975 and entered discussions with ERDA about how and when to begin research. One *Washington Post* article speculated that the Center's work "could make electric vehicles practical and reduce U.S. dependence on foreign oil."[17]

---

12. David Brand, "It's an Ill Wind, Etc.: Energy Crisis May Be Good For Windmills," *Wall Street Journal,* Jan. 11, 1974.
13. "NASA Sells World's Largest Wind Turbine to Hawaiian Electric," *Lewis News,* vol. 25, No. 3 (Feb. 5, 1988), p. 1.
14. "Ohio to Seek Solar Institute: NASA's Plum Brook Facility Could be Considered as Site," *Sandusky Register,* Dec. 10, 1975.
15. "Village to Get Its Power from Sun," *Los Angeles Times,* May 17, 1978.
16. Howard W. Douglass, chief of the Lewis chemical energy division, to Robert E. English, regarding "Hybrid Vehicle Planning Exercise," June 24, 1975, Box 214, Division 8000, NASA Glenn archives.
17. "Battery Power," *Washington Post,* Feb. 16, 1978.

Electric urban vehicle at Lewis Research Center. (NASA Glenn Research Center [NASA GRC].)

The changing focus of the Center's activities prompted rumors—emphatically denied—that it would become part of the ERDA. This even resulted in one report that asked "Should the Agency Continue an Aeronautical Propulsion Program at Lewis?"[18] The Lewis engineers responded by unionizing, and in December 1974, instead of joining the American Federation of Government Employees, they created the new Engineers and Scientists Association and became part of the International Federation of Professional and Technical Engineers. They also looked for a way to return to the roots and the expertise of the Center—engine research. They found their major new mission in the growing national need to develop more efficient engines for commercial aircraft. The new emphasis on energy-efficient aircraft, unlike the ERDA projects, promised to keep Lewis firmly in NASA's fold.[19] Moreover, it brought high visibility to the aeronautics side of NASA, long overshadowed by the enormous budgets and prestige of the space program.

---

18. "How Should NASA Conduct Research and Technology in Aeronautical Propulsion, Should the Agency Continue an Aeronautical Propulsion Program at LeRC?" draft report, Box 260, Division 8000, NASA Glenn archives.
19. Dawson, *Engines and Innovation*.

From 1973 to 1976, according to Donald Nored, the head of the Lewis ACEE programs, "there was much action at Lewis, at Headquarters, and within the propulsion industry addressing fuel conservation."[20] Preliminary studies explored technology concepts that improved efficiency. At the time, Nored remembered, a strong national need fostered a climate that was favorable and aggressive in its support of research. Concepts, ideas, and programs were plentiful, but Nored explained that the genesis of many of the original ideas was blurred because of the frequent interaction and "synergism in the activities." Nonetheless, the period from 1973 to 1976 demonstrates the early articulation of ideas that eventually led to Lewis's three main projects in the ACEE program—Advanced Turboprops, a new energy-efficient engine, and engine performance improvements and deterioration studies.

National need prompted Lewis engineers to begin their fuel efficiency studies 3 years before ACEE's inception. In April 1973, 6 months before the OPEC oil embargo, the Energy Trends and Alternative Fuels study began at NASA Headquarters, with Lewis and Ames assisting. The goal was to identify alternative fuel studies and project fuel usage requirements in the future. Abe Silverstein, Lewis's former powerful Director, chaired the Alternative Aircraft Fuels Committee. By the end of the year, discussions centered on recommending programs more specifically for aircraft fuel conservation and conventional and unconventional modifications to aircraft engines.

In January 1974, a steering committee performed design studies, explored new fuel-conservation technologies, and suggested modification to existing engines. Its work concluded 1 month later, with a plan to establish an Energy-Conservative Aircraft Propulsion Technology program, an ambitious, 9-year plan, accompanied by a funding request of $136 million. By April of that year, cost-benefit analyses were presented to Headquarters. A main component of the project was a new energy-efficient engine, which some speculated would be 30 percent more efficient than existing engines and could possibly be ready for service by 1985. This project eventually evolved into the Energy Efficient Engine program.

The Advanced Turboprop had its origins in an American Institute of Aeronautics and Astronautics (AIAA) workshop in March 1974. After much discussion, the participants agreed that a 15-percent fuel savings

20. Nored, "ACEE Propulsion Background," Jan. 14, 1980, unpublished report, Box 277, Division 8000, NASA Glenn archives.

was possible. The Engine Component Improvement program traced its beginnings to summer 1974, when Lewis engineers awarded a contract to American Airlines, allowing them to examine the airlines' records to begin looking at how its JT8D and JT3D engines deteriorated over time. These records provided early clues as to the extent and cause of the performance decline of the engines. Pratt & Whitney also entered into a contract with Lewis to investigate similar issues and in January 1975 offered its findings on performance deterioration for its current engines. It was at this time that the Kramer Committee took the lead in coordinating NASA's efforts in aircraft fuel conservation, working to establish one central program to organize these activities. Kramer, according to Nored, was "very successful in guidance of the program . . . through the various Headquarters/OMB/industry advisory board pitfalls that can squelch a new start."[21] The ACEE program was underway, and Lewis engineers were anxious and enthusiastic about their three aircraft propulsion projects.

Curing Sick Engines—Engine Component Improvement

It was Raymond Colladay's responsibility to establish the three ACEE propulsion projects at Lewis Research Center. Having started his career at Lewis in 1969, he moved to NASA Headquarters in 1979 to become the Deputy Associate Administrator of the Office of Aeronautics and Space Technology, and then head of DARPA in 1985. Colladay recalled that, at the time he was helping to develop the ACEE program, it was an easy sell to Congress. "The general tenor of Congress and the country as a whole was focused on energy efficiency," and "therefore the Congress was pretty receptive to NASA trying to do what it could in research for energy efficiency." The biggest hurdle was the Office of Management and Budget (OMB). Ideologically, its concern was the proper role of Government in a research and development enterprise. The OMB did not want NASA developing applications for the aircraft industry. While this was not a problem for the majority of the ACEE programs, Colladay said, "the area that caused them the greatest concern was the ECI program because it was component improvements in existing engines, existing aircraft engines."[22]

21. Ibid.
22. Interview with Colladay by Bowles, July 21, 2008.

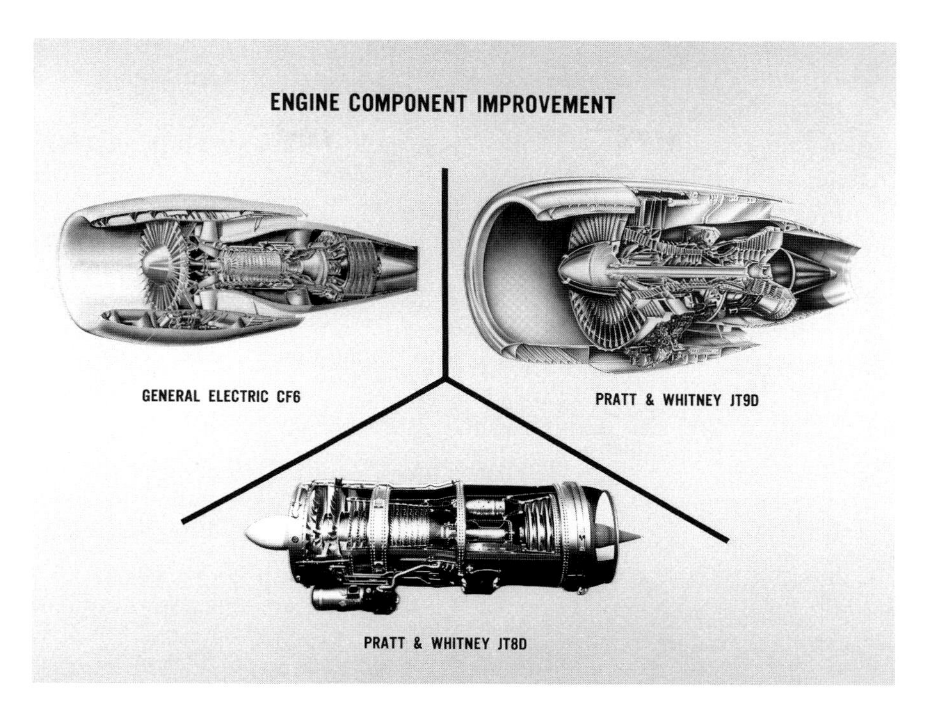

Three of the engines studied in the Engine Component Improvement (ECI) project. The ECI engineers' mission was to improve various components on existing engines that were most likely to wear and decrease fuel efficiency. (NASA Glenn Research Center [NASA GRC].)

The Engine Component Improvement project was unique among all the ACEE programs in that it was expected to return quick results. While other projects looked to incorporate fuel savings advances over 10 to 15 years, ECI aimed to incorporate new technologies within 5 years. The project did not call for revolutionary advances or fundamental changes to existing airplanes. Instead, the mission of the ECI engineers was to improve the components on existing engines that were most likely to wear and decrease fuel efficiency. Pratt & Whitney Aircraft and General Electric manufactured most of the commercial aircraft engines in the United States in the 1970s, and both of these companies collaborated closely with Lewis Research Center on the ACEE project. According to the ECI statement of work, written in December 1976, the main objectives of the program were to "(1) develop performance improvement and retention concepts which will be incorporated into new production of the existing engines by the 1980-1982 time period and which would have a fuel savings goal of 5 percent over the life of these engines, and (2) to provide additional technology which can be used to minimize the performance degradation of current and future engines."[23]

In 1976, four jet engines that were responsible for powering all commercial aviation in the United States. These engines consumed 10-billion gallons of fuel per year.[24] The ECI project focused specifically on developing fuel-saving techniques for the JT9D, JT8D, and CF6 engines. It ignored the JT3D, the fourth major engine, because most industry analysts believed it would not be produced in the future. Introduced in 1964, the Pratt & Whitney JT8D engine was a "phenomenal success" and at its height of popularity flew 12,000 aircraft of different types.[25] Two years later, Pratt & Whitney introduced the JT9D engine, often referred to as opening a "new era in commercial aviation," because it was the first high-bypass engine to power a wide-body aircraft. It was first installed on the Boeing 747 Jumbo Jet, and Pan American placed the first order for this new jet in April 1966.[26] The CF6, a General Electric engine first introduced in 1971, was used on the DC-10 and became the cornerstone of its wide-body engine business for more than 30 years.

The organizations involved in the ECI program read like a who's who of the airlines industry in America at the time. Beginning in February 1977, NASA awarded the two major contracts to General Electric and Pratt & Whitney.[27] Because these companies stood to increase their sales significantly thanks to these NASA advances, a cost recoupment clause was included in their contracts. They were to pay to the U.S. Treasury a 10-percent return on every sale of one of these improved engine components, which was how the ACEE administrators persuaded the OMB to let

---

23. ACEE Engine Component Improvement Statement of Work, Dec. 7, 1976, Box 208, Division 8000, NASA Glenn archives.
24. Joseph A. Ziemianski, "Project Plan, Engine Component Improvement Program," Feb. 1976, Box 244, Division 8000, NASA Glenn archives.
25. R.V. Garvin, *Starting Something Big: The Commercial Emergence of GE Aircraft* Engines (Reston, VA: American Institute of Aeronautics and Astronautics, 1998), p. 126.
26. Klaus Hünecke, *Jet Engines: Fundamentals of Theory, Design, and Operation* (London: Zenith Press, 2005), p. 15.
27. General Electric, "A Proposal for CF6 Jet Engine Component Improvement Program," Sept. 8, 1976, Box 251, Division 8000, NASA Glenn archives. Pratt & Whitney Aircraft, "Technical Proposal for the JT8D and JT9D Jet Engine Component Improvement Program," Sept. 8, 1976, Box 251, Division 8000, NASA Glenn archives.

Pan Am-Boeing 747 flying in 1975. It was one of the main types of aircraft used to test and incorporate ACEE fuel-saving technology. (NASA Glenn Research Center [NASA GRC].)

them go ahead with the ECI project. Every engine that went into active service and had a component traceable to ECI triggered this recoupment. Colladay recalled, "It was a bigger headache than any money it derived, and NASA never saw the money anyway, it went into the Treasury."[28]

General Electric and Pratt & Whitney then established subcontracts with American Airlines, Trans World Airlines, United Airlines, Douglas Aircraft, and Boeing. In addition, Lewis Research Center also contracted with Pan American (for an international route analysis) and Eastern Airlines (for domestic analysis of the technology) to review the program independently and provide ongoing assessments for 30 months.[29] All of these contracts called for three specific tasks: feasibility analysis, development and evaluation in ground test facilities, and in-service and flight-testing. According to Colladay, the reason for the inclusion of essentially all the major airlines

---

28. Interview with Colladay by Bowles, July 21, 2008.

29. Robert J. Antl to manager of the Engine Component Improvement Project Office, regarding negotiations of airline support contracts with Eastern Airlines and Pan American World Airways, Mar. 2, 1977, Box 208, Division 8000, NASA Glenn archives.

in the United States was to "generate a broad base of support" and ensure the highest probability that the ECI technology would be rapidly retrofitted into existing engines or incorporated into new engine builds.[30]

Although getting this broad base of support was important, it did generate some problems—most notably in the relationship between General Electric and Pratt & Whitney. Though within the ECI program they worked together with NASA, in the real world, General Electric and Pratt & Whitney were fierce competitors. Theirs was a historic rivalry. After World War II, Pratt & Whitney dominated in the commercial aircraft engine market, while General Electric was more closely aligned with the military. However, their spheres of influence shifted over time, and by 1977, Pratt & Whitney began losing ground to General Electric in the commercial market. This set the stage, in the early 1980s, for what some have called the "great engine war" between the two companies.[31]

Because of this, the collaboration was sometimes difficult. Pratt & Whitney thought there were "major problem areas" with their relationship. Nored, head of the NASA Energy Conservative Engines Office, admitted that the office was having "extreme difficulty" with Pratt & Whitney and said, "I think they are suffering a corporate reaction to the increasing competition by GE (JT9D vs. CF6)." Both of these engines were scheduled to be improved within ECI. Nored thought the company was nervous about the Freedom of Information Act and as a result wanted to classify all of its research as proprietary. Pratt & Whitney also, in his opinion, sought more and more governmental support to "augment their technology in ways that can influence immediate sales." In accepting this assistance, the company had to learn how to work in the much more open governmental research atmosphere, and sometimes this included being bedmates with chief rivals. For example, General Electric had expressed no concerns about sharing proprietary information, and Nored concluded that Pratt & Whitney needed to "bite the bullet."[32] The program continued despite its often-stated concerns.

---

30. Colladay, "Suggested Response to Congressional Inquiry on Cost Recoupment on the Engine Component Improvement Program," Box 244, Division 8000, NASA Glenn archives.
31. Robert W. Drewes, *The Air Force and the Great Engine War* (Washington, DC: National Defense University Press, 1987), p. 79.
32. Nored to director of aeronautics, Feb. 8, 1979, Box 244, Division 8000, NASA Glenn archives.

There were two main thrusts to ECI—Performance Improvement and Engine Diagnostics. The Performance Improvement section began with a feasibility study to examine a variety of concepts and to prove which one might offer the highest fuel-savings results for the airlines industry. The study looked at the development of an analytical procedure to determine possible concepts, the identification and categorization of concepts, preliminary concept screening, and detailed concept screening. Engineers evaluated 95 concepts for the Pratt & Whitney engines and another 58 concepts for the General Electric engine. The job of the airline industry was to "assess the desirability and practicality of each concept."[33] The concepts were evaluated on two main criteria—technical and economic factors. Technical factors included performance, weight, maintenance, fuel-savings potential, material compatibility, development time, and technical risk, while economic factors included fuel prices, engine cost, production levels, operating costs, return on investment, and life expectancy. Using these criteria, the 153 initial concepts were quickly reduced to 18 and 29, respectively. They were then reviewed in greater detail by NASA and the airlines, which identified 16 concepts that could meet their goals.

The content of these projects can be broken into several important areas. The first was leak reduction. An aircraft engine is similar to an air pump in that it moves air from in front of it to the back. By adding energy to it, the speed of the air moving through the exhaust is faster than what originally came through the inlet. Any air leak in this system caused it to be inefficient, just like an air pump leak. ECI engineers looked for areas in which engine seals could be improved to reduce this leakage. A second major area for improvement was in aerodynamics: ECI engineers developed improved designs of the compressor and turbines. A third area was ceramic coatings on components, which was important because it reduced the necessity of cooling holes and both increased efficiency and reduced manufacturing costs.

Specifically, the 16 projects, and their related engine types, were as follows. For the JT8D, they included an improved high-pressure turbine air seal, high-pressure turbine blade, and a trenched tip high-pressure compressor. JT9D improvements for the high-pressure turbine included a

---

33. Pan American World Airways contract with NASA Lewis Research Center, Feb. 28, 1977, Box 208, Division 8000, NASA Glenn archives.

ceramic outer seal, a thermal barrier coating, active clearance control, and new fan technology. CF6 improvements were a new fan, a front mount for the engine, a short core exhaust nozzle, improved aerodynamics for the high-pressure turbine, a roundness control for the turbine, and active clearance controls for the turbine. There were two other aircraft-related projects: a nacelle drag reduction for the DC-9 and compressor bleed reduction for the DC-10. The ECI Performance Improvement program was significant thanks to its success after only a few years of research, testing, and development. According to Jeffrey Ethell, "By 1982 most of the improved components were flying and saving fuel, giving the companies involved a firm leg up in the commercial aircraft marketplace, where they were being challenged by foreign competition."[34]

The Engine Diagnostics program focused on analyzing and testing the JT9D and CF6 engines.[35] Pan Am engineers considered this to be the "most significant work" of the ECI program. An often-used logo for the Engine Diagnostics program was an engine with a human face, frowning, tongue sticking out, and arms clasped over its midsection. A country doctor hunched over it, tools sticking out of his pockets, examining an x-ray machine, diagnosing a way to "cure the sick engine." While just a caricature, it did simplistically convey the fundamental goals of this program. The engine "illnesses" were the performance losses they experienced as their flight hours increased. The "doctors" were the Lewis engineers, whose job was to determine the mechanical sources of these problems and recommend ways to "cure" the sick machines. Their recommendations could keep existing engines healthy and help to prevent the deterioration of future engines.[36]

One known problem with these or any type of engines was that over time, various components begin to deteriorate because of operational stresses, which included combustors that warped because of continual fluctuation in temperatures from hot to cold, compressor blades whose tips wore down over time, seals that began to leak hot gases, and turbine blades

34. Ethell, *Fuel Economy in Aviation* (Washington, DC: NASA SP-462, 1983), p. 20.
35. Engine Component Improvement Program, Monthly Project Management Report, Sept. 30, 1977, Box 206, Division 8000, NASA Glenn archives.
36. "NASA JT9d Engine Diagnostics Program," Sept. 14, 1977, Box 225, Division 8000, NASA Glenn archives.

eroding from high temperatures. Other types of damage could occur when foreign objects such as stones or dust entered the engines on the runway and caused dents, breaks, or scratches to the fan blades. The engines were durable and could typically fly for 10,000 hours before they needed major maintenance, but during that time, the engine slowly became less and less fuel efficient because of small degradations that did not compromise the safety of the aircraft. Furthermore, the major maintenance sessions never restored the engines to their original levels of fuel efficiency. Pan American engineers said that prior to the ECI Engine Diagnostics program, "engine deterioration had been largely a matter of educated guessing, speculation, and hand-waving."[37] This deterioration became the focus of the Engine Diagnostics program, and engineers estimated that by preventing these wear-and-tear issues, aircraft would become more fuel efficient.

Engine Diagnostics engineers from NASA, General Electric, and Pratt & Whitney began their work by evaluating the existing data on performance deterioration from the airline industry and engine manufacturers. The data included in-flight recordings, ground-test data, and information on how frequently various parts were repaired and replaced. Additional data, needed on the JT9D and CF6, were obtained though special monitoring devices, as well as analysis gained from a complete teardown and evaluation of the engines. Special ground tests were developed to experiment with short- and long-term performance deterioration. These ground tests helped engineers simulate operating conditions to determine the sources of component deterioration. From the data they collected, they identified certain components whose failure rates could be improved upon.[38]

One concern, raised by Pratt & Whitney, was that the deterioration information on its engines was being used by its competitors. Its company slogan was "Dependable Engines," and extensive publications as to how they deteriorated over time was, in its opinion, damaging to its reputation.[39]

---

37. N.B. Andersen and L.H. Allen, Engine Component Improvement Program, Airline Evaluation, Dec. 19, 1980, Report of Meeting No. 10, Box 208, Division 8000, NASA Glenn archives.

38. ACEE Engine Component Improvement Statement of Work, Dec. 7, 1976, Box 208, Division 8000, NASA Glenn archives.

39. Pratt & Whitney Aircraft, "Summary of the Proposal for the JT8D and JT9D Engine Component Improvement Program," Sept. 8, 1976, Box 251, Division 8000, NASA Glenn archives.

Specifically, the company had evidence that Rolls-Royce, a British engine competitor, used the ECI deterioration data from the JT9D and CF6 engines in its then-current marketing campaign, demonstrating the superiority of Rolls-Royce engines. A 1979 letter from Pratt & Whitney's legal team to NASA expressed concerns that the ECI program would "adversely affect" its marketing and future sales potential. The team wanted NASA to change its dissemination policy for technical reports to protect the Pratt & Whitney "marketing position" for its engines.[40] NASA responded that this was an unintended consequence of the ECI program and the effort to improve engines for the United States airlines industry. Furthermore, according to NASA, Pratt & Whitney's role in the program had been voluntary and had the "full backing and support of P&W management." NASA officials had informed the companies at the outset that comparisons between the engines would be made, and both Pratt & Whitney and General Electric "realized the consequences of entering into the program and accepted them."[41]

The independent outside evaluations by Pan American and Eastern Airlines were an important part of the ECI project. The independent reports by Pan American World Airways are especially revealing. The reports were based upon 10 meetings held during the project in which NASA, General Electric, and Pratt & Whitney representatives summarized their work for the Pan America review committee. The first meeting, held at John F. Kennedy Airport in March 1977, was a get-acquainted session for the various participants to discuss early concepts, directions, and goals for the project.[42] By the sixth meeting, in September 1978, Pan American was expressing serious concerns, characterizing the program as "disappointing" and criticizing the ECI engineers for taking a "very conservative approach," rather than a "considerably more aggressive" one. "We are also greatly concerned that the manufacturers appear to be losing sight of the basic objective of this program," Pan American concluded at the

---

40. Robert M. Gaines, Assistant Counsel for Pratt & Whitney, Feb. 13, 1979, Box 244, Division 8000, NASA Glenn archives.

41. Nored, response to Pratt & Whitney on reporting marketing-sensitive data on NASA ECI program, Feb. 8, 1979, Box 244, Division 8000, NASA Glenn archives.

42. N.B. Andersen and J.G. Borger, Engine Component Improvement Program, Airline Evaluation, Mar. 22, 1977, Report of Meeting No. 1, Box 208, Division 8000, NASA Glenn archives.

time.[43] By the end of the program, Pan American engineers saw significant areas of improvement and at one of the final reviews offered substantial praise to the program, saying, "In spite of what may have been interpreted as high critical comments during various review presentations"the program has resulted in "important knowledge" and was "quite successful."[44]

In fact, the ECI project was one of the more successful of the ACEE programs, for several reasons. The first reason was the speed with which improvements were incorporated onto commercial aircraft—many of the projects findings found their way into commercial aircraft engines before other ACEE programs even had their first test flights. John E. McAulay, the head of the ECI Performance Improvement project, presented the positive results of the project's work at the January 1980 Aerospace Sciences Meeting, just 3 years after the program began. While it "has already provided significant potential for reductions in the fuel consumed by the commercial air transport fleet," he said he was optimistic that even greater savings were possible through their ongoing studies.[45] By March 1980, the ECI engineers had produced 20 technical papers, 21 contractor reports, 4 technical memorandums, 6 conference publications, and 8 journal and magazine articles.[46]

Second, the organizations that benefited most from the project were very enthusiastic about the results when ECI ended. In 1980, Harry C. Stonecipher (General Electric vice president and general manger) wrote to John McCarthy (Director of Lewis Research Center) to highlight the program's value to his company, writing that it generated a "wealth of knowledge" and that its main beneficiaries were the airline industry. He estimated the savings of this "invaluable" research at a reduction of 10 gallons of fuel for the CF6 engine for each hour of flight. Stonecipher concluded, "We at General Electric want you to be aware of the benefits this program has provided, and the tremendous potential for the years ahead."[47]

---

43. Andersen and Borger, Engine Component Improvement Program, Airline Evaluation, Oct. 31, 1978, Report of Meeting No. 6, Box 208, Division 8000, NASA Glenn archives.
44. Ibid.
45. John E. McAulay, "Engine Component Improvement Program—Performance Improvement," Jan. 1980, Aerospace Sciences Meeting, AIAA Paper 80-0223.
46. "Engine Component Improvement Report Status," Mar. 1980, Box 224, Division 8000, NASA Glenn archives.
47. Harry C. Stonecipher to John McCarthy, May 5, 1980, Box 260, Division 8000, NASA Glenn archives.

Third, the ECI program helped to maintain the competitive advantage of the entire commercial aircraft industry. For example, in February 1980, Boeing executives approached NASA to ask if they could disclose results of the ECI program to foreign airlines, because in order to sell new American aircraft in the international marketplace, the company needed to show its more advanced understanding of engine deterioration and how to improve engine performance. NASA agreed with Boeing's request and stated, "In order to meet the challenge presented by international competition, it is appropriate that the U.S. aircraft industry use the technology generated in the ECI program to maintain its dominant position in the marketplace."[48]As Roger Bilstein wrote, "Research results were so positive and so rapidly adaptable that new airliners in the early 1980s like the Boeing 767 and McDonnell Douglas MD-80 series used engines that incorporated many such innovations."[49] Though the fuel-efficiency rewards were never intended to be as high as in other ACEE programs (including the Energy Efficient Engine), ECI was successful in achieving a significant fuel reduction of roughly 5 percent, exactly what its engineers projected at the onset of the program.

The Frontiers of Engine Technology—The Energy Efficient Engine

In the early 1980s, the aircraft industry had endured numerous difficulties, including reduced profitability, increasing fuel costs, higher worker wages, political pressures with deregulation, and increasing worldwide competition. Many once-dominant airlines were fighting for their survival, including Pan Am. Pratt & Whitney and General Electric, two of the leading U.S. engine manufacturers, were "cutting each others' throats, and prices," and experiencing increasing difficulties competing in the world market against the British government-owned Rolls-Royce.[50] But according to one 1983 report, despite these problems, the "airline industry in the years ahead

48. David J. Poferl, manager of the Advanced Propulsion Systems Office, memorandum for the record regarding a teleconference with Boeing's Dick Martin, Feb. 25, 1980, Box 244, Division 8000, NASA Glenn archives.
49. Roger Bilstein, *Orders of Magnitude: A History of the NACA and NASA, 1915–1990* (Washington, DC: NASA SP-4406, 1989), p. 117.
50. Howard Banks, "A Job Well Done," *Forbes* (Oct. 10, 1983), p. 146.

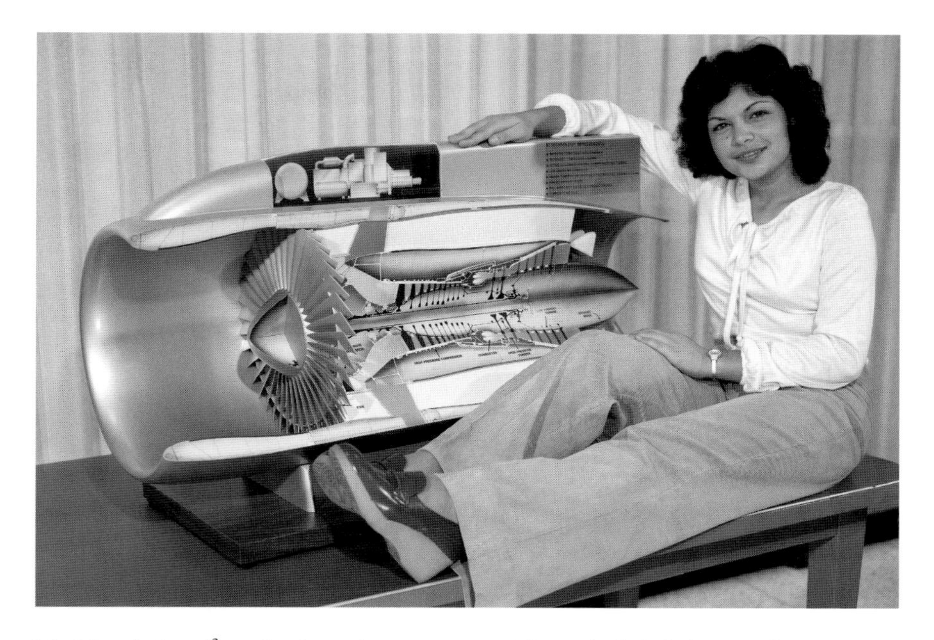

Model of the E^3 technology improvements. These included improved component aerodynamics, improved compressor loading, active clearance control, low emissions combustor, and higher-temperature materials. (NASA Glenn Research Center [NASA GRC].)

looks a bit rosier." One major reason cited for this optimism was a "less noticed effort" that involved the redesign of the aircraft engine itself.[51] This was another ACEE project managed by Lewis Research Center, known as the Energy Efficient Engine. As *Forbes* magazine reported, E^3 was a "NASA success story . . . thoroughly overshadowed by the glamorous space programs."[52]

Given their intense competition, Pratt & Whitney and General Electric were strange bedfellows, but they continued this relationship in the E^3 project. Each organization had ideas about how to improve fuel efficiency for aircraft engines, but neither was willing to accept the risk, in both time and money, to develop these ideas on its own. NASA stepped in to assume the majority of the risk, providing $90 million to each company, with a promise that each would invest $10 million of its own. This program had

51. Skip Derra, "Joint R&D Program Improves Aircraft Engine Performance," *Industrial Research & Development* (Nov. 1983), p. 79.
52. Banks, "A Job Well Done," p. 146.

GE Energy Efficient Engine (June 16, 1983). (NASA Glenn Research Center [NASA GRC].)

several main goals: to reduce fuel consumption by 12 percent, decrease operating costs by 5 percent, meet FAA noise regulations, and conform to proposed EPA emission standards. Additional goals included guidelines for minimum takeoff thrust and a safe and rugged engine with a 10-percent weight reduction.[53] The engines used for benchmarking fuel efficiency were the same ones used for the ECI studies—the Pratt & Whitney JT9D and the General Electric CF6. Also as in the ECI program, these two prime contractors worked with the airlines to discuss engine design options. These included Boeing, Douglas, and Lockheed. Eastern Airlines and Pan American served as additional advisers and contributed operational experience.

The program was managed by Carl Ciepluch at Lewis (as well as Raymond Colladay for a time), Ray Bucy at General Electric, and W.B. Gardner at Pratt & Whitney. Bucy was extremely enthusiastic about this program, saying that the E^3 program was "guiding the future of aircraft engines."[54] Fuel-efficient aircraft were very complex technological systems that required extensive and costly research, he believed, but the rewards would be well worth the investment. Bucy hoped the resulting engine would save 1-million gallons of fuel per year for each aircraft flying commercially. Gardner even thought that the program would surpass its expectations "beyond the program goal."[55]

That goal was to have a new turbofan engine ready for commercial use by the late 1980s or early 1990s. A turbojet derived its power and thrust entirely from the combustion and exhaust of its burning fuel.[56] A turbofan is also a turbojet, but it has an extra set of rotating, propeller-like blades, positioned ahead of the engine core. The air from the fan goes partly through the engine core, and the remainder flows around the outside the engine. The "bypass ratio" is the ratio of air flowing around the engine to the air flowing through it. When this ratio is either 4 or 5 to 1, the engine is referred to as a "high-bypass engine." The high-bypass turbofans were more efficient than were either the turbojets or the earlier low-bypass

---

53. General Electric, "Original Work Plan for Energy Efficient Engine Component Development & Integration Program," Apr. 28, 1978, Box 239, Division 8000, NASA Glenn archives.

54. R.W. Bucy, "Progress in the Development of Energy Efficient Engine Components," ASME paper 82-GT-275, Box 181, Division 8000, NASA Glenn archives.

55. W.B. Gardner, W. Hannah, and D.E. Gray, "Interim Review of the Energy Efficient Engine E^3 Program," ASME paper 82-GT-271, Box 181, Division 8000, NASA Glenn archives.

56. For more on the "turbojet revolution," see: Edward W. Constant, *The Origins of the Turbojet Revolution* (Baltimore: Johns Hopkins University Press, 1980).

engines developed in the 1950s and 1960s. However, by the 1970s, the high-bypass engines promised greater potential for application to wide-body commercial aircraft, although one of their main problems was their environmental impact, in terms of noise and emissions.[57] The potential of the high-bypass turbofan engine was the E^3 program's main goal.

The idea for incorporating high-bypass engines into the existing commercial airline fleet began in 1974. Two investigations—the "Study of Turbofan Engines Designed for Low Energy Consumption," led by General Electric, and the "Study of Unconventional Aircraft Engines Designed for Low Energy Consumption," led by Pratt & Whitney—demonstrated a great deal of promise. Both studies suggested to NASA the importance of new high-bypass engines. But, as was so often the case, "the cost of such programs . . . [was] enormous," and the time required to accomplish it was at least a decade.[58] To make the development more feasible for industry, the report suggested a continued joint effort led by NASA, with the results made available to all airlines and engine manufacturers. Without governmental support, such an open research atmosphere would have been impossible. "Results from these studies," wrote Colladay and Neil Saunders, "indicated enough promise to initiate the EEE project."[59]

In the E^3 program, both General Electric and Pratt & Whitney were given the task of building a new turbofan engine. But the idea was not for them to build a commercial-ready engine. The E^3 engine was to be used primarily for testing and proof of fuel-efficient concepts. The new technological components included a compressor, fan, turbine-gas-path improvements, structural advances, and improved blading and clearance control. Although the contractors had the same goal, they approached their work within E^3 differently.[60] Pratt & Whitney engineers took a

---

57. Jack L. Kerrebrock, "Aircraft Energy Efficiency Program Status," Subcommittee on Transportation, Aviation and Materials, Feb. 17, 1982, Box 234, Division 8000, NASA Glenn archives.
58. American Airlines and Pratt & Whitney, "Technology of Fuel Consumption Performance Retention," unpublished report, Feb. 1974, Box 181, Division 8000, NASA Glenn archives.
59. Colladay and Neil Saunders, "Project Plan, Energy Efficient Engine Program," June 1977, Box 272, Division 8000, NASA Glenn archives.
60. R.V. Garvin, *Starting Something Big: The Commercial Emergence of GE Aircraft Engines* (Reston, VA: AIAA, 1998), p. 167.

Energy Efficient High Pressure Compressor Rig (April 10, 1984). (NASA Glenn Research Center [NASA GRC].)

"component" strategy and concentrated on developing a high-pressure turbine that could be operated with a lower temperature of hot gas to improve efficiency. General Electric proceeded with a more comprehensive approach, researching the best way to integrate a new fan, high-pressure compressor, and low-pressure turbine. According to Jeffrey Ethell, the freedom that the contractors had was important: "The 'clean sheet' opportunity . . . gave both companies the chance to leave their normal line of evolutionary development and leap forward into high-risk . . . areas to research and aggressively push the frontiers of technology."[61]

Along with these two prime contractors, there were subcontracts with major commercial airframe manufacturers. Boeing, Douglas, and Lockheed provided expertise in areas related to airplane mission definitions and engine and airframe integration. Just as in the ECI program, Eastern Airlines and Pan American also provided ongoing evaluation of the results from the perspective of the airlines. NASA also planned

61. Ethell, *Fuel Economy in Aviation,* p. 30.

to use its own in-house technological advances and other contractors to support specific program needs. NASA never intended to develop a new engine as a product. This was a project for the engine manufacturers to achieve after NASA assisted with the proof of concepts. Elements of the ECI program such as improved fans, seals, and mixers were incorporated into the E^3 program, and the E^3 engineers were also able to apply results from the ECI Engine Diagnostic program to improve engine performance.[62]

A first step in the E^3 program was to identify risk factors that might potentially cause the new engine to fail. In an April 1976 letter from James Kramer, Director of the ACEE office, to Donald Nored, the chief of the Energy Conservative Engines Office at Lewis, Kramer asked that the Center perform a "risk assessment of the total E^3 program."[63] With a list of potential failures in hand, the Center could better understand the implication on schedules, cost, and program success. A separate action plan could then be put in place to reduce these risks. Two months later, Nored and Lewis completed the risk assessment. "By nature," wrote Nored, "this is a high risk program, as is true of most advanced technology programs, and there is no way to make it a safe bet."[64] The best way to minimize risk, according to Nored, was to use multiple contractors who were supplied with adequate funds. Both General Electric and Pratt & Whitney took on separate areas of risk that were unique challenges to their approaches and engines. With both companies involved, Nored believed "at least one-half or greater of the stated goal" would be achieved.

As the program got underway, one important advance was a computer control system known as a full authority digital electronics control (FADEC). It could monitor and control 10 engine parameters at the same time and communicate information to a pilot. Sensors were known to be one of the least reliable of all engine components. The FADEC system was able to compensate for this problem in case of failure by modeling what the engine should be doing at any given time during a flight. If the sensor

62. Colladay and Neil Saunders, "Project Plan, Energy Efficient Engine Program," June 1977, Box 272, Division 8000, NASA Glenn archives.
63. Kramer to Nored, Apr. 20, 1976, Box 181, Division 8000, NASA Glenn archives.
64. Nored to Kramer, June 1, 1976, Box 181, Division 8000, NASA Glenn archives.

failed, then the FADEC, based on its model, could tell the various engine components what they should be doing.[65]

In 1982, budget reductions caused "program redirection" for the E[3] project. According to Cecil C. Rosen, the manager of the Lewis propulsion office, this meant changes for both General Electric and Pratt & Whitney in how they planned to complete the project. General Electric proceeded with its core engine test and suspended work on emissions testing and an update for a flight propulsion system. For Pratt & Whitney, the redirection meant a continued focus on component technology as opposed to an overall engine system evaluation. The main concern with this plan was that it provided more funding for Pratt & Whitney than General Electric because it had "much farther to go in its component technology efforts." Rosen hoped this "unequal funding," which went against the original spirit of the E[3] program, would be acceptable.[66]

General Electric completed the program with a great deal of success and as early as 1983 was being called the "world's most fuel-efficient and best-performing turbofan engine."[67] Bucy, the Program Manager at General Electric, called it "one of the most successful programs on an all-new engine in years."[68] While the low-pressure turbine was a difficult challenge from an aerodynamic perspective, it achieved the desired parameters laid out by NASA at the start of the program to define success. There is a 13-percent improvement in fuel efficiency over the CF6-style engine, which was 1 percent better than required. GE immediately began to incorporate the new technology into its latest engine designs, including the CF6-80E, the latest engine for the Airbus A330, and the GE90 engine for the Boeing 777.[69] The GE90 first made headlines in 1991 because it "pushed the edge of technology," not only because it was more efficient, but also because it used another ACEE project. It became the only engine

---

65. Derra, "Joint R&D Program Improves Aircraft Engine Performance," *Industrial Research & Development* (Nov. 1983), p. 80.

66. Cecil C. Rosen, III, to Lewis chief of transport propulsion, Feb. 22, 1982, Box 181, Division 8000, NASA Glenn archives.

67. "E[3] Considered World Leader in Fuel Efficiency," *Design News* (Dec. 5, 1983).

68. Bucy, quoted in "E[3] Considered World Leader in Fuel Efficiency," *Design News* (Dec. 5, 1983).

69. Garvin, *Starting Something Big: The Commercial Emergence of GE Aircraft Engines* (Reston, VA: AIAA, 1998), p. 167.

to use composite fan blades, making it 800 pounds lighter, with a 3.5-percent fuel savings. It also had a cleaner burn, producing 60 percent less nitrous oxide, and was quieter. Though it was a larger engine, its engineers believed that the wind whistling over the landing gear would produce more noise than the engine. As Christopher D. Clayton, the manager of the GE90 technical programs said, "It will give us a much more efficient engine. That's the real purpose of it."[70] The 777 now flies with an engine based directly upon the one developed through the efforts of the E^3 ACEE program.

Pratt & Whitney also had success with its energy efficient engine technology, though at a slower pace. In 1988, it reported that the "efficiency trends show a steady increase"[71] with the E^3 technology. But the company still had research to perform to enable it to realize the gains for "tomorrow's engine." These successes were finally realized in 2007, when it launched the new energy-efficient Geared Turbofan as the engine for the Mitsubishi Regional Jet. This was a 70- to 90-seat passenger aircraft, and Mitsubishi planned to purchase 5,000 of them over the coming 20 years. The technology for this engine could be directly traced back to Pratt & Whitney's participation in the ACEE program.[72]

These favorable results of the E^3 program, as well as the achievements of the ECI program, resulted in enthusiasm for ACEE. In 1979, Colladay said, "This early success in the first of the ACEE Program elements to near completion is certain to continue as more of the advanced concepts are put into production."[73] However, this "continued certainty" was seriously threatened in 1980 with a new presidency on the horizon. Unlike ECI, which returned such fast and positive results, the other ACEE programs required a longer window to develop and prove their technologies, and their engineers required a commitment of time and money from the United States Government to ensure that their research continued. Just 3 years after the entire program began, there were serious concerns not

---

70. Christopher D. Clayton, as quoted in Dick Rawe, "GE90: Future Power Plant," *The Cincinnati Post,* Dec. 17, 1991.

71. Pratt & Whitney, "Advancement in Turbofan Technology . . . NASA's Role in the Future," 1988, Box 121, Division 8000, NASA Glenn archives.

72. Rob Spiegel, "Pratt-Whitney Builds Energy-Efficient Engine for Mitsubishi," *Design News*, Dec. 10, 2007.

73. Colladay, "Engine Component Improvement Program," Sept. 24, 1979, Box 244, Division 8000, NASA Glenn archives.

only for the future of ACEE, but for the future of all aeronautics activities at NASA. For the ACEE participants, the question was: Would the Government terminate such a vital fuel-efficiency program to the Nation early, when it had already had such success with its shorter-term projects like the Engine Component Improvement? For NASA, the question was even more dire: Would the Agency be allowed to continue its work in aeronautics?

Chapter 4:

Aeronautics Wars at NASA

On September 11, 1980, 2 months before the presidential election, Ronald Reagan wrote a letter to Gen. Clifton von Kann, the senior vice president at the Air Transport Association of America. In it, he outlined the aeronautical objectives of his potential presidency, as well as his criticisms of the Carter era. While not questioning the importance of aviation to the economic and military strength of the Nation, Reagan was highly critical of ongoing programs. "I am deeply concerned about the state of aeronautical research and development," he wrote, using as an example the "alarming" slowdown in aircraft exports to other nations, an industry the United States had once dominated. He identified energy efficient aircraft as one critical aviation issue facing the Nation, and of these efforts, he wrote, "Our technological base is languishing." Reagan promised von Kann, "The trends must be reversed. And I am committed to do just that."[1]

Soon after Reagan assumed the presidency, a conservative think tank that played a major role in shaping the philosophy of his Administration made the shocking assertion that the NASA aeronautics program was actually eroding the country's leadership in aviation. The group, the Heritage Foundation, said, "*The program should be abolished.*"[2] The new Reagan Presidential Administration began to seriously consider taking aeronautics away from NASA and letting industry assume the primary role in research. Richard Wagner, the head of the Laminar Flow Control program at Langley, said, when "Reagan came into office . . . we didn't know whether

1. Ronald Reagan to Gen. Clifton von Kann, Sept. 11, 1980, Box 234, Division 8000, NASA Glenn archives.

2. Italics in original. Eugene J. McAllister, ed., *Agenda for Progress: Examining Federal Spending* (Washington, DC: Heritage Foundation, 1981), pp. 171–172.

President Ronald Reagan gets a laugh from NASA officials in Mission Control when he jokingly asks astronauts Joe Engle and Richard Truly if they could stop by Washington en route to their California landing. To his right is NASA Administrator James Beggs (November 13, 1981). (NASA Headquarters—Greatest Images of NASA [NASA HQ GRIN].)

we were going to stay alive or not . . . it was a struggle."[3] It was an even greater struggle for his colleagues at Lewis Research Center, as the entire base was threatened with closure. The conflict between NASA and other Government agencies had started slowly in 1979, with general disagreements over language in an ACEE report by the Government Accounting Office (GAO). Conflict escalated over budget reductions in 1980 and subsequent cuts for ACEE. But by 1981, the clash became a full-on fight for survival as the Reagan Administration pushed for the closure of Lewis Research Center, the elimination of aeronautics from all of NASA, and the termination of over 1,000 aeronautics jobs. These were the aeronautics wars.

ACEE Battles with the GAO

The roots of this struggle predated Reagan's inauguration, originating with the controversial 1979 GAO report. For much of NASA's history,

3. "NASA Aircraft Energy Efficiency Program Marked for Elimination," *Aerospace Daily,* vol. 113, No. 3 (Jan. 6, 1982), p. 17.

aeronautics programs had never required close oversight because of the low levels at which they were traditionally funded. The large budget and greater visibility of ACEE suddenly brought it unwanted attention. In June 1979, the heads of several key NASA programs were asked to comment on "where we're going in aeronautics." Donald Nored, Director of the Lewis ACEE program, responded by saying that this is the type of question that is "perhaps best explored in a leisurely retreat" but complied by answering in a four-page letter. In it, he said that NASA should be responsive to major national needs and should focus on "break-through, innovative, novel, high risk, and high payoff" programs. Nored, who was far from an unthinking cheerleader, was critical of some aspects of the NASA aeronautical program. He said that there seemed to be a "hodge-podge" of activities, with NASA trying to do too many things and responding to industry in areas that were too evolutionary and incremental. NASA's aeronautics should be, according to Nored, "more revolutionary."

Nored's final suggestion addressed what he believed to be the most important issue facing aeronautics in 1979 and the future—fuel. "The problem of fuel," he said, "is an overriding problem to all other technical issues in the field of aeronautics." Nothing was more important than solving the technological issues represented by rising fuel costs because it threatened the airline industry and the continuation of American prosperity. Nored believed that ACEE was a good beginning, but that even more needed to be done. He advocated for "continued vigorous support" of this and other related fuel efficiency activities, and he pleaded for what he called "agency *urgency*."[4]

Nored's views were almost entirely discounted in a GAO report on ACEE. In August 1979, 2 months after Nored made his suggestions, the GAO released a draft review that was highly critical of the ACEE project. It was the first in a series of reviews the GAO planned for all of NASA's aeronautical projects. Since ACEE had the greatest visibility and importance among all of them, it received the first of the governmental reviews. Under the direction of the House Committee on Science and Technology, the review's goal was to "recommend potential program options for replacing ACEE," and the GAO went on a 6-month fact-finding mission

4. Emphasis in original. Nored to the NASA director of aeronautics, June 4, 1979, Box 238, Division 8000, NASA Glenn archives.

in 1979 to Lewis and Langley Research Centers, 3 airframe companies, and 2 jet engine companies.[5]

The first observation made by the report was that it was "unclear" if ACEE—which had, after all, only been operational for a few years—would achieve its objectives, although it found that NASA had some "limited technology successes to date." All of this should have been a likely observation, since the 10-year program was still, for the most part, in its beginning phases and was attempting some risky and revolutionary aeronautical research. The report did indicate one of the main reasons the results were unclear—funding. The programs with the highest fuel-savings potential—the Advanced Turboprop, Laminar Flow Control, and Composite Primary Aircraft Structures—were threatened because neither Congress nor the Carter Administration would commit to funding. The report concluded that "the cumulative affect [*sic*] of these uncertainties highlights why the meeting of ACEE objectives is currently very unclear."[6]

The GAO's analysis of specific ACEE programs was also critical. Of the Lewis projects, it had little positive to say. The report stated that the Engine Component Improvement project was "falling short" of its performance goal. The Energy Efficient Engine was too new to evaluate, and its chances of meeting goals were "unknown." Likewise, the GAO admitted that it was "too early to say" if the Advanced Turboprop would be a success, but that it was 3 years behind schedule. The Langley ACEE programs did not fare much better. The Energy Efficient Transport was criticized because it appeared to the GAO that only Douglas Aircraft would be able to integrate the new fuel-saving technologies, and not Boeing or Lockheed. The GAO called the prospects for the Laminar Flow Control program "uncertain," believed that NASA was further away than originally thought to achieving its goals, and claimed the program was 4 years behind schedule. Finally, the GAO criticized the Composite Primary Aircraft Structures program for failing to develop a composite wing or fuselage, underestimating costs, and not foreseeing the hazardous potential effects of carbon fiber releases into the environment. The GAO concluded that the

---

5. General Accounting Office, "Preliminary Draft of a Proposed Report, Review of NASA's Aircraft Energy Efficiency Project," Aug. 1979, Box 182, Division 8000, NASA Glenn archives, p. 9.
6. Ibid., p. 16.

President Jimmy Carter presents the National Space Club's Goddard Memorial Trophy to NASA Administrator Robert A. Frosch on behalf of the team that planned and executed the Voyager mission. (NASA Headquarters—Greatest Images of NASA [NASA HQ GRIN].)

composites program would achieve "dramatically less" fuel savings than originally projected.[7]

On January 24, 1980, NASA Administrator Robert Frosch responded vigorously to the draft. Frosch wrote to J.H. Stolarow, the GAO Procurement Director, that after reviewing the report with officials at Langley and Lewis, "We are very concerned about the negative tone of the report and its implications regarding the value of the NASA Aircraft Energy Efficiency (ACEE) Program." Frosch criticized the reviewers for basing evaluations of the program upon schedules set up in 1975, before the program began. More significantly, he said, the report trivialized the major advances that ACEE had already achieved. Frosch put the full weight of his support behind the program and described it as a "significant contributor" to the overall aviation research and technology program in the United States. He praised the Government and industry team for its cooperation and added that the results would have a "major influence on transport aircraft of the future."[8]

---

7. Ibid., pp. 18, 26, 34, 42, 44, and 52.

8. Robert A. Frosch to J.H. Stolarow, Jan. 24, 1980, Box 182, Division 8000, NASA Glenn archives.

ACEE managers at Lewis, Langley, and Headquarters wrote a more detailed response to the GAO and fought to have its conclusions changed before the GAO released the final report. They argued that in general, the GAO presented a "distorted view" that, if left unchanged, would create the "false impression that the program has been less than successful," jeopardizing future funding for the program and leading essentially to a self-fulfilling prophecy.[9] NASA produced a several-page document that provided a thorough review on how large portions text of the report should be altered to better reflect the realities of the ACEE program.

Their efforts were successful in persuading the GAO to craft a much more positive document. In the final report, "A Look at NASA's Aircraft Energy Efficiency Program" (July 1980), the GAO explained its reversal of language and opinion, saying that in light of NASA's concerns, it "carefully reevaluated its presentation and made appropriate adjustments where it might be construed that the tone was unnecessarily negative or the data misleading."[10] For example, the first sentences of the original draft chapter on the ACEE status read: "The prospects of ACEE achieving its objectives are unclear. Technical readiness dates are being slipped."[11] This tone was significantly changed in the final published report, which said: "The ACEE program, which is in its 5th funding year, has experienced some technological successes which will be applied on new and derivative airplanes built in the early 1980s. Examples are improved engine components, lighter airframe components, and improved wings."[12]

Changes were also made to specific program reviews. John Klineberg, a member of the founding ACEE task force and eventual Lewis Center Director, said that in the original report, the GAO treated the turboprop project unfairly. He called the reviewers ignorant of the project's "inherent uncertainties," because from the start, it was considered one of the more

---

9. NASA response reprinted in, "A Look at NASA's Aircraft Energy Efficiency Program," by the Comptroller General of the United States, July 28, 1980, Box 182, Division 8000, NASA Glenn archives, p. 65.

10. Ibid., p. iv.

11. General Accounting Office, "Preliminary Draft of a Proposed Report, Review of NASA's Aircraft Energy Efficiency Project," p. 11.

12. "A Look at NASA's Aircraft Energy Efficiency Program," by the Comptroller General of the United States, July 28, 1980, Box 182, Division 8000, NASA Glenn archives, p. 5.

risky ACEE programs.[13] Lewis project managers prevailed in persuading the GAO to cast a more favorable light on the turboprop. In the draft, the GAO argued: "The Task Force Report shows that in 1975 there was considerable disagreement on the ultimate likelihood of a turboprop engine being used on commercial airliners."[14] In the final publication, the GAO amended the sentence to read: "The possible use of turboprop engines on 1995 commercial aircraft is still uncertain, but has gained support since 1975."[15] These editorial adjustments demonstrated the effectiveness of project managers working to improve public and governmental understanding of the project. They also highlight the political skills often necessary to ensure technological success, or the perception of success, at NASA.

In August 1980, 1 month after reading the final report, Walter B. Olstad, the Acting Associate Administrator for Aeronautics and Space Technology, felt as if the battle had been won. He said that upon final review, the report "fairly stated" the ACEE progress. He was also pleased to report that in almost every area that NASA expressed objections, the GAO made appropriate changes. Olstad wrote, "while a great deal of our responses to the draft versions of the GAO ACEE report may have sounded negative . . . [we] appreciate the opportunities afforded during its preparation to make substantive inputs."[16]

The battle exemplified by the NASA and GAO conflict was not unusual. Institutional conflict is more the norm than the exception. In an article about NASA during the Reagan years, political scientist Lyn Ragsdale wrote that conflict between Congress, the Presidency, and NASA occurred often because they operated within a system of separate institutions that all shared a power mitigated through checks and balances. "In order to circumvent such conflict," according to Ragsdale, "officials in one or more institutions must be willing to invest political capital to raise

---

13. Klineberg to NASA Headquarters, Dec. 21, 1979, Box 182, Division 8000, NASA Glenn archives.
14. General Accounting Office, "Preliminary Draft of a Proposed Report, Review of NASA's Aircraft Energy Efficiency Project," p. 37.
15. "A Look at NASA's Aircraft Energy Efficiency Program," by the Comptroller General of the United States, July 28, 1980, Box 182, Division 8000, NASA Glenn archives, p. 45.
16. Walter B. Olstad to Office of Inspector General, Aug. 28, 1980, Box 182, Division 8000, NASA Glenn archives.

public awareness."[17] The political fights for ACEE did not end with the GAO conflict. Instead, they intensified as ACEE managers and NASA leaders fought to raise awareness not only of the importance of fuel-efficiency aviation programs, but of NASA's role in aeronautics itself.

Woods Hole Versus the Heritage Foundation

In summer 1980 (just as the GAO report was coming out), NASA's aeronautical leaders organized an independent review of its entire aeronautics program by the Aeronautics and Space Engineering Board (ASEB). ASEB's members included some of the most influential people in United States aviation and aerospace history. Chaired by Neil Armstrong, the 24-member board included representatives from NASA (the former Johnson Space Center director), industry (vice presidents or technology directors at Douglas, Pan American, Sikorsky, United Technologies, Grumman Aerospace, Boeing, General Electric, and Lockheed), and academia (noted aeronautics professors from Stanford, MIT, and the California Institute of Technology). To discuss the state of aeronautics at NASA, the board held a workshop that ran from July 27 to August 2, 1980, at the National Academy of Sciences Study Center in Woods Hole, MA. Sixty experts were divided into five panel sessions. The chairman of the workshop, H. Guyford Stever, called it an "arduous and exhilarating week-long effort" to examine every facet of NASA and its role in aeronautics.[18] Its conclusions became known as the "Woods Hole Plan."

The Woods Hole Plan was unveiled to the public in a document titled *NASA's Role in Aeronautics: A Workshop*. ASEB agreed that there had been a long and important relationship between the aviation industry and the Government, which started with the NACA and continued through NASA. This had been a positive relationship that had strengthened the American industry, helping it position itself better for competition in the world market. However, this historical strength had faced significant threats in

---

17. Lyn Ragsdale, "The U.S. Space Program in the Reagan and Bush Years," *Spaceflight and the Myth of Presidential Leadership*, Roger D. Launius and Howard E. McCurdy, eds. (Urbana: University of Illinois Press, 1997), p. 140.

18. H. Guyford Stever, Chairman of the Aeronautics and Space Engineering Board Workshop, to Robert W. Rummel, Chairman of the Aeronautics and Space Engineering Board, Jan. 16, 1981, *NASA's Role in Aeronautics: A Workshop*, vol. 1, Box 260, Division 8000, NASA Glenn archives, p. vii.

recent years. The ASEB members at Woods Hole emphasized that there was an "urgent need" to counter these economic, social, political, and technological challenges facing the United States in aviation. The United States had lost 20 percent of the world's aircraft market to European competition during the previous several years, in part because European governments collectively endorsed a plan to displace the United States as the world's aviation leader. As a result of this government support, these European nations were able to cut into the dominance of the American transport market. To counter the ongoing European threat, ASEB called for greater U.S. governmental intervention and assistance, not less, in order to equip the aviation industry to compete.

Above all, the ASEB aviation experts said it was the worldwide concern about the cost and availability of fuel that would potentially have the most important influence over the future of aviation. It pointed to "dramatic improvements" already attained in fuel efficiency through improved aerodynamics, materials, and propulsion, most importantly through the ACEE program. They concluded, "World leadership in aeronautics will be achieved, in all probability, by the nation or nations that seize the initiative and move such technologies from their present research status . . . [to build] more efficient aircraft." The only way to achieve this and reverse the "erosion of momentum" of the American aeronautical technology was to "clarify and strengthen NASA's role in aeronautics."[19] NASA was extremely pleased with ASEB's findings and believed that they would be most valuable should any criticism of its aeronautics program emerge. NASA would not have to wait long to confront the critics.

While the Woods Hole group was writing its findings, Republican campaign strategists began defining the shape of a future Reagan presidency. Although the election was still 2 months away, Edwin Meese, the Chief of Reagan's campaign staff, said he wanted a low-visibility effort as far as planning making plans for a Reagan presidency that would not detract from the campaign. One of the key groups assisting in this planning was the Heritage Foundation, the nonprofit conservative think tank established in 1973.[20] In October 1980, a spokesman for the organization said it

19. Ibid., pp. v–vi and 4–5.

20. Adam Clymer, "Staff Quietly Plans for a Reagan Presidency," *New York Times*, Sept. 14, 1980.

would establish a "comprehensive game plan for implementing conservative policy goals under vigorous White House leadership."[21] This included, in part, reducing the budget, balancing it, and restoring "moral values." It became what was called a "blueprint for the construction of a conservative government." Meese told reporters that he would be relying heavily on it.

Fall 1980 was a time of great uncertainty for NASA. John Noble Wilford, a *New York Times* reporter, wrote in September 1980 about NASA's launch of a Delta rocket at Cape Canaveral to place a weather satellite into orbit. Wilford said this launch might represent the "death of the National Aeronautics and Space Administration as we know it."[22] It was a difficult time for the Agency. There were numerous difficulties with the still unlaunched Space Shuttle. Of its first 17 scheduled missions, only 2 were defined by NASA, with the Pentagon taking significant control of the others. The Department of Defense was investing in rocketry and satellites, and NASA was becoming more of a service agency that launched spacecraft for other nations. Budgets were being cut, and NASA was getting little support from either of the United States presidential candidates. NASA's Administrator, Robert Frosch, announced his retirement in October 1980, to be effective on Inauguration Day, January 20, 1981.[23] NASA employees eagerly awaited the results of the presidential election and wondered how it would shape their future. They would soon find out.

Twelve days after Reagan won, the Heritage Foundation published a report, the *Mandate for Leadership*, which became the blueprint for the new presidency. Called by the *Los Angeles Times* a "quick strike a week after Reagan's election," the report began the process of dismantling 48 years of New Deal liberal policies.[24] It included such suggestions as abolishing the Department of Energy, reassigning most of the functions of the Environmental Protection Agency to the states or other Federal agencies, and increasing the defense budget by $20 billion. While some called it the

---

21. William Endicott, "'Think Tank' Drawing Up Plans to Achieve Conservative Goals in Reagan Presidency," *Los Angeles Times,* Oct. 4, 1980.
22. John Noble Wilford, "Others Tread on NASA's Piece of Sky," *New York Times,* Sept. 14, 1980.
23."NASA Chief Intends to Resign His Post on Inauguration Day," *Wall Street Journal,* Oct. 7, 1980.
24. "Conservative Think Tank Moves Into Capitol Spotlight," *Los Angeles Times,* Dec. 21, 1980.

most complete report on government ever written, one observer said, "The political fall-out . . . will be great. Opposition will be savage."[25] Meese's strong endorsement of it was in part responsible for it appearing on the *Washington Post*'s bestseller list for 3 weeks in 1981. It became the bible of the Reagan Administration.[26]

The report argued that the Government should no longer play a role in the commercialization of technology. It contended that Government's commercialization endeavors had been expanding in recent years, and while there were certain areas where this was necessary—such as weapons labs, uranium enrichment, and other areas of nuclear research—on the whole, these activities should stop. Aviation was not on the list of appropriate areas for commercialization. The report concluded, "Generally it should not be the function of the Federal government to involve itself with the commercialization of technology."[27] While ACEE was not explicitly a commercialization project, it did push the lines of development farther than most NASA aeronautics programs had in the past, and so it became vulnerable to cancellation by the Reagan Administration. Reagan's science adviser from 1981 to 1985, George A. Keyworth, explained that it was a "new era" for American industry, and specifically for industrial R&D, an area that it would offer new opportunities for industry to exercise its inventiveness and ingenuity, while at the same time challenging it to accept new roles and to fund research previously supported by Government on its own.[28]

A philosophy explicitly opposed to governmental support of aeronautics research was more completely articulated in another Heritage Foundation report, the *Agenda for Progress*. It said that NASA was spending $500 million each year for research related to civil and military aeronautical technology and that it could find "no good justification for the federal government to spend money on this program." The foundation also criticized NASA for diverting skilled engineers away from

---

25. Joanne Omang, "The Heritage Report: Getting the Government Right with Reagan," *Washington Post,* Nov. 16, 1980.
26. Donald E. Abelson, *A Capitol Idea: Think Tanks and US Foreign Policy* (Montreal: McGill-Queen's University Press, 2006), p. 34.
27. Charles L. Heatherly, ed., *Mandate for Leadership: Policy Management in a Conservative Administration* (Washington, DC: Heritage Foundation, 1981), p. 235.
28. George A. Keyworth, II, "The Federal Role of R&D," *Research Management* (Jan. 1987), pp. 7–9.

profitable aeronautical ventures in industry and toward careers that supported a particular political agenda. The Heritage Foundation claimed that the continuation of the existing aeronautical policy would eventually "erode our leadership, not strengthen it." The report concluded by saying that taxpayers should not bear the burden for this program. The solution was for the aircraft companies to "finance their own research," and for the NASA aeronautics program to be "*abolished*."[29]

Although aeronautics engineers at NASA and in industry were extremely disappointed—some were enraged—they were not entirely taken off-guard. NASA countered the Heritage Foundation assertions with the Woods Hole report. One NASA official wrote to Donald Nored that the "Woods Hole Plan is a strong endorsement of NASA's program, at an opportune time."[30] NASA's administrators used the report to raise public awareness and secure aeronautical support from Congress. In February 1981, NASA's Acting Administrator, Alan Lovelace, wrote letters and provided copies of the Woods Hole report to members of all the congressional committees and subcommittees associated with aeronautics. Olstad, the Acting Associate Administrator for Aeronautics and Space Technology, initiated his own related campaign as well. He wrote, "We have been concerned for some time that the practices and guidelines used by NASA to carry out its aeronautical programs are not generally well understood."[31] He hoped the report would clarify that aeronautical mission with a concise public statement about this NASA responsibility and its importance to the Nation. Olstad then sent the report to NASA's center directors, including Donald P. Hearth at Langley and John F. McCarthy at Lewis, and provided them with copies of the ASEB report.[32]

Despite these advocacy efforts, on February 5, 1980, the Reagan Administration announced plans to slash the NASA budget by 9 percent.

---

29. Emphasis in original. Eugene J. McAllister, ed., *Agenda for Progress: Examining Federal Spending* (Washington, DC: Heritage Foundation, 1981), pp. 171–172.
30. Lynn Anderson to Nored, Feb. 23, 1981, Box 260, Division 8000, NASA Glenn archives.
31. Lovelace to Edwin "Jake" Garn, Chairman of the Subcommittee on HUD-Independent Agencies, Feb. 9, 1981, Box 260, Division 8000, NASA Glenn archives.
32. Olstad to center directors, Feb. 17, 1981, Box 260, Division 8000, NASA Glenn archives.

The *Washington Post* reported that these cuts "took NASA's top management by surprise."[33] While the official hit list naming which programs would be targeted for reduction had not been established, the aeronautics engineers knew they were in jeopardy. With NASA as a whole fighting for survival, the aeronautics budget threatened, and the ACEE managers deeply concerned about the continuation of their program, Nored decided NASA needed to focus on marketing. In January 1981, Nored produced a document establishing advocacy guidelines for the aeronautics programs in NASA, writing that the "effective advocacy or 'selling' of new programs is essential to the health of [NASA]." His document was used by the aeronautics directorate personnel within NASA to conduct an "effective 'marketing' campaign which will eventually lead to approval of their proposed new programs by Congress."[34] Walter Stewart, the NASA Lewis Director of Aeronautics, called this emphasis on advocacy and marketing "vital to our well being."[35]

It was also very timely. In March 1981, NASA's Deputy Administrator, Alan Lovelace, gave an impassioned plea on Capitol Hill at the NASA budget hearings. The No. 1 problem America faced was the national economy, he said, and aviation's role was to serve as a model for reestablishing worldwide economic leadership. He outlined some of his major concerns: for the first time, a major U.S. airline purchased a fleet of foreign-made aircraft, the French-made Airbus had begun outselling the most advanced U.S. transport by a 3 to 1 ratio throughout the world, and enrollment in aeronautics courses at colleges and universities was at an all-time low. In the midst of these threats, Lovelace said NASA faced curtailment of its aviation programs with a new governmental philosophy regarding the aviation industry: "let them go it alone."

Lovelace explained that the situation was dire and said, "Because I am not happy enough to sing," he would paraphrase the lines of a vintage Bob Dylan song. The pertinent lyrics were, "My friends the message is blowing

---

33."Planet Exploration Dwindles in 'Hit List' on NASA's Budget," *Washington Post,* Feb. 5, 1981.

34. Nored, "Guidelines for Advocacy of New Programs in the Aeronautics Directorate," Jan. 1981, Box 121, Division 8000, NASA Glenn archives.

35. Walter L. Stewart to R.J. Weber, Feb. 2, 1981, Box 238, Division 8000, NASA Glenn archives.

in the wind; the message is blowing in the wind." Lovelace testified that he believed the Woods Hole report stated well the reasons for support of NASA's aeronautics program. He reiterated that American leadership in aviation has been sustained and cultivated by the work of the NACA and NASA, in collaboration with industry. NASA, in his opinion, and in the opinion of experts from Government, industry, and academia, needed to be able to continue its aeronautical research to help stimulate the airlines industry and strengthen the American economy. The model had been successful for decades, and there appeared no reason to change it fundamentally during a period of intense international threat to American leadership. Lovelace spoke directly to Ronald Reagan when he said, "My message, then, Mr. President, can be summarized by saying simply: Let us keep that beacon brightly lit and let us supply the fuel to do it."[36]

Lovelace's plea had little effect. The resulting budget, presented after the hearings, was disappointing. Congress cut NASA's funding by $219 million. While support for programs such as the Space Shuttle remained unchanged, aeronautics programs lost $33 million in funding as compared with the previous year. Of these, ACEE saw program reductions of $7 million, including a $5.5-million reduction for the Energy Efficient Engine. Though funding was maintained for Laminar Flow Control, the budget postponed important ground evaluations for 2 years. Lovelace concluded that the effect of these reductions "will be significant," but that they are not "crippling."[37] The most crippling threats were still to come.

Fighting to Save Lewis and Aeronautics at NASA

Significant problems remained on the horizon for NASA's aeronautics efforts even after the budget reduction debate in March 1981. The OMB, under direction from the Reagan White House, continued pressing a plan that would fundamentally change NASA. In response, a variety of influential individuals from the Department of Defense and Congress

36. Lovelace, hearings before the Subcommittee on Science, Technology, and Space, Serial No. 97-29, Part I, Mar. 10, 1981.

37. Lovelace, statement before the Subcommittee on Space Science and Applications, Mar. 31, 1980, Box 234, Division 8000, NASA Glenn archives.

fought alongside NASA to prevent the OMB from dissecting the Agency and amputating its aeronautics arm.

In November 1981, Secretary of Defense Casper Weinberger became aware of the plan to eliminate aeronautical research at NASA. The specifics of the plan, according to Weinberger, would "result in the closing of Lewis Research Center," as well as the loss of over 1,000 aeronautics jobs at NASA. While NASA would be removed from civil aeronautics work, it would continue to support the development of military aircraft. This was at a time when the Government had the green light to expand significantly the Nation's defense, and Weinberger became concerned that the closure of Lewis and the changes to NASA at this critical moment would weaken the development efforts of the B-1B Bomber (and other military programs). So Weinberger wrote a letter to OMB Director David Stockman saying, "I am deeply concerned that the proposed reductions will adversely impact [these] programs, and are not consistent with DOD needs."[38] Before any action to close Lewis or to eliminate the aeronautics program was taken, Weinberger said, the Defense Department should review the consequences of these actions.

Weinberger was known as such a staunch cost-cutter in Washington that he was often called "Cap the Knife."[39] But this was one instance when he fought to keep a program intact. Weinberger had his Undersecretary of Defense, Richard D. DeLauer, immediately contact NASA Administrator James Beggs. In a letter dated November 30, 1981, DeLauer told Beggs that the OMB was "proposing major reductions" in the 1983 budget for the "aeronautics technology program." These reductions would change the landscape of NASA itself, including the "closing of Lewis Research Center" and also the "substantial reductions in aeronautics activities" at Ames and Langley Research Centers.[40] Thirteen hundred other aeronautics personnel would also be eliminated throughout NASA. DeLauer said many of the advanced Department of Defense programs were "critically dependent on a vital and productive NASA aeronautics program."

---

38. Casper Weinberger to David Stockman, Nov. 1981, Box 234, Division 8000, NASA Glenn archives.

39. Weinberger, *In the Arena: A Memoir of the 20th Century* (Washington, DC: Regnery, 2001), p. 191.

40. Richard D. DeLaurer to James Beggs, Nov. 30, 1981, Box 234, Division 8000, NASA Glenn archives.

James M. Beggs was sworn in as NASA's sixth Administrator at a White House ceremony July 10, 1981. Officiating was Vice President George Bush. At center is Beggs's wife, Mary. Beggs was previously an executive vice president and director of General Dynamic Corp. (July 10, 1981). (NASA Headquarters—Greatest Images of NASA [NASA HQ GRIN].)

He then made the essential argument for keeping NASA involved in civil aircraft work: "We should not lose sight of the fact that manufacture of civil aircraft contributes not only to the economy, but also the maintenance of the industrial base which is so important to DOD under surge conditions." (It is interesting to note that NASA's Administrator, Daniel Goldin, from 1992 to 2001 removed NASA from the DOD connections that represented such important support for the aeronautical program during the lean years. According to Joseph Chambers, "After NASA cut the cords, the DOD labs established their own specialists and forgot who NASA was. That situation exists today—in spades.")[41]

NASA also garnered the support of the Army. On December 1, 1981, Beggs received a letter from Jay R. Sculley, the Assistant Secretary of the

41. Chambers, correspondence with Bowles, Mar. 28, 2009.

Army. Sculley again confirmed the closure rumors and told Beggs that, in his view, the relationship with NASA was "essential to the Army to in furthering its R&D programs."[42] The expertise that was resident at the various NASA Centers was as unique and vital as the aeronautical facilities under their control. If these were to disappear, the result would be a dramatic increase in funding requests by the Army to offset those NASA reductions. The net effect would be the expenditure of more money. From the Army's perspective, this was a counterintuitive and damaging step for the OMB to make.

This view was also supported by Dan Glickman, a Congressman from Kansas and the Chairman of the Subcommittee on Transportation, Aviation, and Materials. In November 1981, he invited aviation industry leaders to a hearing to discuss "The First 'A'" in NASA, which of course was "Aeronautics." The hearing was held December 8, 1981, and its goal was to document the historic role of Federal support of aeronautics to determine if funding should continue. He told his invitees that "some in the Reagan Administration have suggested that the NASA Aeronautics program be drastically curtailed."[43] This was, according to Glickman, a "radical departure," and all the consequences and ramifications needed to be understood. He sent letters to all the major commercial airframe and engine manufactures in the United States, including General Electric.

Though in the "First 'A'" hearings NASA fought to retain a central piece of its heritage, the story did not merit enough attention to be covered by the Nation's major newspapers. The only NASA news reports during this period discussed the status of the Space Shuttle and the hopes of some enthusiasts to send a probe to Halley's Comet. But the hearings did draw the attention of the aeronautics industry and politicians in Cleveland, OH, the home of the endangered Lewis Research Center. Mary Rose Oakar, who represented Lewis's congressional district, fought Capitol Hill for the preservation of 2,700 jobs at Lewis and the millions of dollars of tax revenue the Center generated for Ohio. She invited President Reagan to come to Lewis to see for himself how vital a laboratory it was, describing it as a "beacon of the highest form of technology research."[44]

---

42. J.R. Sculley to Beggs, Dec. 1, 1981, NASA Glenn archives.

43. Dan Glickman to Thomas Donohue, general manager, Aircraft Engine Group, General Electric, Box 234, Division 8000, NASA Glenn archives.

44. Statement of Mary Rose Oakar, Dec. 8, 1981, Box 234, Division 8000, NASA Glenn archives.

Thomas Donohue, the general manager from the General Electric aircraft engines group, provided a historical overview of the important aeronautical work NASA and the NACA performed for the Nation and called for the Government to keep this tradition alive.[45] Other aircraft and engine manufacturers provided similar supporting comments, and after the "The First 'A'" hearings, Glickman sent letters to the CEOs of each of these companies. In his communication with Jack Welch, at General Electric, he praised Donohue's testimony and urged Welch to write to President Reagan and lend his endorsement that aeronautics deserved to remain within NASA.[46]

One of the Heritage Foundation's main arguments was that aeronautics was a "mature" technology and therefore did not need active Government-supported research. ACEE program proponents refuted this stance. Brian Rowe, a General Electric senior vice president, wrote a response to this question by Victor H. Reis, Assistant Director, Office of Science & Technology Policy: "Is aeronautics a stagnant technology?" He said, quite simply, "No!" Rowe firmly believed that with continued research, the aeronautics industry would see a rate of progress over the next 20 years similar to that of the previous 40. He used as a specific example the important gains still to be realized in fuel efficiency, and he projected that the "the fuel consumed per passenger on an inaugural flight of an airliner in the year 2002 will be 40% to 50% less than that of the first revenue service of the new Boeing 767 later this year."[47] Aeronautics, in his view, was not a mature technology, and ACEE was spearheading many of the developments that would enable the United States to maintain its worldwide aeronautical leadership.

The results of these "First 'A'" hearings were discussed at the critical February 1982 budget hearings for NASA's fiscal 1983 funding. Glickman said the hearings results demonstrated unanimous support in rejecting the Reagan Administration's plan to shift the burden of aeronautical research to industry and eliminate NASA from this work.[48]

---

45. Statement of Donohue, Dec. 8, 1981, Box 234, Division 8000, NASA Glenn archives.
46. Dan Glickman to John F. Welch, Jr., Dec. 8, 1981, NASA Glenn archives.
47. Brian H. Rowe to Victor H. Reis, assistant director, Office of Science & Technology Policy, Apr. 23, 1982, Box 234, Division 8000, NASA Glenn archives.
48. Glickman, hearing before the Subcommittee on Transportation Aviation and Materials, Feb. 17, 1982, Box 234, Division 8000, NASA Glenn archives.

Despite the groundswell of support, the OMB pushed forward with the plan to slash aeronautics. A headline in *Defense Daily* stated that budget cuts were "Forcing NASA to Close Lewis Research Center," and many in Washington saw its closing as fait accompli.[49] Likewise, the headlines of an *Aerospace Daily* article read, "NASA Aircraft Energy Efficiency Program Marked for Elimination."[50] Though the program had been achieving impressive gains at both the Langley and Lewis Centers, the funding cuts proposed by the OMB threatened ACEE because it was a program that directly benefited industry, and this went against the grain of the Reagan philosophy. But the announcements of the demise of Lewis and ACEE were premature. Though funding cuts were a significant loss for aeronautics in 1983, it was not an across-the-board termination of the program. The insistence of the Department of Defense, industry leaders, politicians, and NASA managed to counter the Heritage Foundation's recommendation. The Reagan Administration allowed NASA's aeronautics program and ACEE to limp forward.

NASA responded with an attempt to develop a strategic plan for the future of aeronautics. Hans Mark, the head of Ames Research Center, led the initiative. Jack L. Kerrebrock, the Associate Administrator for Aeronautics and Space Technology, said in February 1982 that the plan would provide long-term goals as well as short-term suggestions for the 1984 fiscal budget.[51] The resulting document, the "Strategic Plan for Aeronautics," included mission statements related to the importance of aeronautics to national policy and an emphasis on maintaining all the existing NASA Research Centers and their areas of expertise.[52] Nowhere was this goal more important than in Cleveland, OH.

In July 1982, Lewis Research Center organized a "Save the Center Committee" with support from the Ohio delegation to Congress and Ohio's Senators, John Glenn and Howard Metzenbaum. It was at this time

---

49. "Budget Cuts Forcing NASA to Close Lewis Research Center," *Defense Daily,* vol. 119, No. 25 (Dec. 9, 1981). Dawson, *Engines and Innovation,* p. 217.
50. "NASA Aircraft Energy Efficiency Program Marked for Elimination," *Aerospace Daily,* vol. 113, No. 3 (Jan. 6, 1982), p. 17.
51. Jack Kerrebrock to directors of Ames, Langley, and Lewis Research Centers, Feb. 19, 1982, Box 215, Division 8000, NASA Glenn archives.
52. "Strategic Plan for Aeronautics," Mar. 1982, Box 234, Division 8000, NASA Glenn archives.

President Ronald Reagan shaking hands with Andrew Stofan, who served as Director of the Lewis Center (April 23, 1986). (NASA Glenn Research Center [NASA GRC].)

that Lewis Center Director John McCarthy stepped down and Andrew Stofan from Headquarters replaced him. Although some were concerned about the timing of this decision, Stofan injected Lewis with a revitalized spirit. Stofan had strong Lewis ties, having served as the director of its very successful launch vehicles program. Upon taking control, he initiated an extensive review of Lewis and started planning not only how to save it, but also how to make it more viable in the future.[53] Through his charisma, confidence, and powers of persuasion, Stofan kept Lewis alive. The strategic plan committee, headed by William "Red" Robbins

53. Andrew Stofan, interview by Bowles, Apr. 13, 2000.

and Joseph Sivo, gave Stofan the task of winning five major new programs for the Center. When he returned from Washington having secured four of them, as well as an indefinite stay of execution for the Center, it was, Robbins said, "a damn miracle."[54] One of the programs Stofan fought to retain funding for was the ACEE Advanced Turboprop Project. Aeronautics across NASA was much weaker than it had been from a budgetary standpoint, but it survived extinction. The two long-range and risky ACEE projects, the Advanced Turboprop (at Lewis) and Laminar Flow Control (at Langley) had opportunities to achieve success and program resolution.

---

54. Red Robbins, as found in Dawson, *Engines and Innovation*, pp. 213–214.

Chapter 5:

Advanced Turboprops and Laminar Flow

A 1987 *Washington Post* headline read, "The Aircraft of the Future Has Propellers on It."[1] To many, this sounded like heralding "the reincarnation of silent movies."[2] Why would an "old technology" ever be chosen over a modern, new, advanced alternative? How could propeller technology ever supplant the turbojet revolution? How could the "jet set mind-set" of corporate executives who demanded the prestige of speed and "image and status with a jet" ever be satisfied with a slow, noisy, propeller-driven aircraft?[3] A *Washington Times* correspondent predicted that the turbojet would not be the propulsion system of the future. Instead, future airline passengers would see more propellers than jets, and if "Star Wars hero Luke Skywalker ever became chairman of a Fortune 500 company, he would replace the corporate jet with a . . . turboprop."[4] It appeared that a turboprop revolution was underway.

The Advanced Turboprop Project was one of the more radical and risky projects in the ACEE program, but it offered some of the highest fuel-efficiency rewards. NASA planners believed that an advanced turboprop could reduce fuel consumption by 20 to 30 percent over existing turbofan engines while maintaining comparable performance and passenger comfort at speeds up to Mach 0.8 and altitudes up to 30,000 feet. These ambitious goals made the turboprop project controversial and challenging. Clifton von Kann succinctly summed up these concerns to

---

1. Martha M. Hamilton, "Firms Give Propellers a New Spin," *Washington Post,* Feb. 8, 1987.
2. Robert J. Serling, "Back to the Future with Propfans," *USAIR* (June 1987).
3. R.S. Stahr, "Oral Report on the RECAT Study Contract at NASA," Apr. 22, 1976, Nored paper, Box 224, NASA Glenn Archives.
4. Hugh Vickery, "Turboprops are Back!" *Washington Times,* Nov. 1, 1984, p. 5B.

Barry Goldwater during his Senate testimony, when he said that of all the proposed projects, "the propeller is the real controversial one."[5]

The Advanced Turboprop was not the only revolutionary, long-range technology in the ACEE program. Some speculated as early as the 1960s that Laminar Flow Control would be a "harbinger of potential revolution in the plane-making business."[6] The Laminar Flow project was based upon an airplane wing that seemed to "breathe" air. When engineers began achieving significant successes with this technology in the early 1960s, they knew they were on the cusp of a major advance. Many wondered if the resulting aircraft with breathable wings would be able to fly for days—and not just hours—without refueling. Or, more realistically, a nonstop flight from New York to Tokyo might be offered to commercial travelers. First flight-tested in 1963, the "air-inhalation system" was considered "the most promising innovation since the jet engine."[7] Because of the Vietnam war, the military suspended further work on this technology, but it was resurrected in the 1970s and became the most promising ACEE project in terms of fuel efficiency.

Lewis Research Center managed the Advanced Turboprop Project, and Langley Research Center headed the Laminar Flow Control program. Although the two NASA ACEE projects had little interaction with each other, they shared some important similarities. First, they represented revolutionary potential in fuel efficiency, with the turboprop promising up to 30 percent and laminar flow up to 40 percent. Second, achieving these gains required commitment from the very conservative American airlines industry to a fundamental and radical new aircraft design and propulsion system. Finally, both programs required a long-term commitment to research, and both had risky and uncertain futures. For these reasons, industry alone would never risk the funds to research their potential, but the Government support through NASA offered an appropriate venue for exploring technology that could have a revolutionary impact on the airlines

---

5. Statement by Clifton F. von Kann to the Senate Committee on Aeronautical and Space Sciences, Sept. 10, 1975, Box 179, Division 8000, NASA Glenn archives.
6. John C. Waugh, "Wings 'Breathe' in Laminar Plane," *Christian Science Monitor,* May 21, 1963.
7. Marvin Miles, "Plane Passes Revolutionary Air Flow Test," *Los Angeles Times,* Aug. 16, 1963. William L. Laurence, "Aviation Landmark: New Design May Permit Aircraft to Stay Aloft for Days," *New York Times,* May 26, 1963.

industry. The questions at the start of the program were: Could NASA engineers achieve success and develop these new fuel-efficient technologies? And, if they could, would the airlines industry accept the challenge and open its arms to incorporate the technology in its new fleet of aircraft?

The Aerodynamicist's Pot of Gold—Laminar Flow Control

Laminar flow control has been an elusive and alluring quest that has tempted aeronautics engineers for nearly 80 years. According to historian James Hansen, "Nothing that aerodynamicists could to do to improve the aerodynamic efficiency of the airplane in the late twentieth century matched the promise of laminar flow control."[8] Richard Wagner, the head of Langley's Laminar Flow Control program, said that of all the ACEE programs, it offered "by far, the biggest payoff."[9] Engineers knew that, if it could be perfected, laminar flow control could improve fuel efficiency by 30 percent or more and decrease drag by 25 percent. Using 2004 estimates, if the United States airlines could reduce drag by just 10 percent and fuel economy by 12 percent, it would result in a savings of $1 billion per year. Albert L. Braslow, who spent his career working in the laminar control field, argued that it was the "only aeronautical technology" that would enable a transport airplane to fly nonstop to any point in the world and to stay aloft for 24 straight hours. He concluded that the incredible fuel savings was the "'pot of gold at the end of the rainbow' for aeronautical researchers."[10] This allusion was perhaps more appropriate than Braslow realized, or would have liked. Though the lure of the rainbow's gold and laminar flow control are undeniable, to this day, neither exists, though the commercial potential for laminar flow remains in sight.

The fundamentals of laminar flow are as follows. When a solid (such as an aircraft wing) moves through air, it encounters friction. The thin layer of air that interacts with the solid's surface is called the boundary layer. Within this layer, two conditions can occur: a laminar condition, where the airflow is uniform in nonintersecting layers, and a turbulent, where

---

8. Hansen, *The Bird Is On the Wing*, p. 203.

9. Interview with Wagner by Bowles, June 30, 2008.

10. Braslow, *A History of Suction-Type Laminar Flow Control with Emphasis on Flight Research* (Washington, DC: NASA *Monographs in Aerospace History* No. 13, 1999), p. 1.

the airflow within the boundary layer is characterized by turbulent eddies that cause additional drag. At lower speeds, conditions are relatively favorable for an aircraft to enjoy the smooth laminar flow over its wing surfaces, tail, and fuselage. But as the speed increases, it becomes more difficult to maintain laminar flow, and a more turbulent boundary layer takes over.[11] For example, a transport plane flying at subsonic speeds spends half of its fuel to maintain normal cruise speeds while attempting to counter the friction and turbulence found in this boundary layer.

Attaining ideal laminar flow is possible in two main ways. Natural laminar flow (also known as "passive") can occur over the leading edge of an airplane's wing by contouring the airfoil to a particular shape. To achieve laminar flow rearward from the leading edge of the wing requires an "active" approach, known as laminar flow control. One of the best approaches is a suction method in which holes or slots in the wing draw some of the boundary layer air through it. Pumps suck the air down through the surface, where ducts vent it back out into the atmosphere. In this way, the wing or airfoil appears to "breathe."

The earliest laminar flow investigations began in the 1930s, when German engineers first developed stability analysis methods. In 1939, Langley engineers began performing wind tunnel tests to study turbulence and laminar flow. The NACA became increasingly interested in studying this phenomenon, and 2 years later, Langley was able to flight-test a B-19 with 17 suction slots in a special test section mounted on one wing panel. During World War II, active laminar flow control work was suspended in order for research to take place on natural laminar flow for aircraft such as the P-51 Mustang, while Germany and Switzerland continued their active approaches. After the war, Langley (aided by the release of confidential German research after World War II to the aeronautics community) returned to suction studies in wind tunnels and provided theoretical support that this approach was indeed possible.[12] The Air Force also became interested in laminar flow and contracted with Northrop Corporation to

---

11. Bill Siuru and John D. Busick, *Future Flight: The Next Generation of Aircraft Technology,* 2nd ed. (Blue Ridge Summit, PA: TAB/AERO, 1994), p. 45.
12. Braslow, Dale L. Burrows, Neal Tetervin, and Fioravante Visconte, *Experimental and Theoretical Studies of Area Suction for the Control of the Laminar Flow Boundary on an NACA 64A010* (Washington, DC: NACA Report 1025, Mar. 30, 1951).

Early laminar flow tests on a blunted 15-degree cone cylinder in free flight at high Reynolds number (July 23, 1956). (NASA Glenn Research Center [NASA GRC].)

investigate suction through slots and holes. The NACA concluded that the main impediment to achieving laminar flow control was the difficulty in creating smooth surfaces on the airplane. Even factors such as bugs or ice crystals could cause the loss of a laminar flow.

Research continued and tremendous optimism surged in the early 1960s over the Air Force's work with laminar flow. In 1963, the *New York Times* announced an "aviation landmark" and a "new aeronautical milestone" with the flight of an X-21, a reconnaissance-bomber research aircraft, and a "revolutionary air-inhalation system."[13] Under the direction of Wener Pfenninger at Northrop, a slot-based laminar flow control system was successfully flight-tested, and some observers called it the most promising development in flight since the jet engine. Even though the Air Force viewed laminar flow as the most "prominent" and "promising" of its leading aerodynamic projects, further research was delayed for another decade.[14]

---

13. William L. Laurence, "Aviation Landmark: New Design May Permit Aircraft to Stay Aloft for Days," *New York Times,* May 26, 1963.
14. Walter J. Boyne, *Beyond the Wild Blue: A History of the United States Air Force* (New York: St. Martin's Griffin, 1998), p. 194.

The center section of each wing of this business jet was modified for tests of laminar flow control (October 15, 1984). (NASA Langley Research Center [NASA LaRC].)

From the mid-1960s to the mid-1970s, laminar flow studies were suspended, in large part because of the commitment of military resources to the war in Vietnam. Also, the low cost of jet fuel completely offset the savings when compared with manufacturing and maintenance costs for aircraft with active laminar flow control.

This economic situation changed with the rise in fuel prices and the end of the war. When NASA began looking at technologies to include in the ACEE program, laminar flow was an early favorite. Langley researchers had resumed studies on it, and in 1973, Albert Braslow wrote a white paper arguing that it had "by far the largest potential for fuel conservation of any discipline."[15] While many were enthusiastic about it, Braslow noted that some managers at NASA Headquarters and Langley were "lukewarm" to the idea. Detractors thought the technological barriers were so

15. Braslow and Allen H. Whitehead, Jr., *Aeronautical Fuel Conservation Possibilities for Advanced Subsonic Transports* (Washington, DC: NASA TM-X-71927, Dec. 20, 1973). Braslow, *A History of Suction-Type Laminar Flow Control,* p. 13.

steep that it would be throwing away limited aeronautics funding to pursue the research.

As fuel costs continued to rise, the promise of laminar flow became more and more attractive. In March 1974, the AIAA held a conference with 91 of its members to discuss aircraft fuel-conservation methods, and they concluded that laminar flow deserved attention. Their ideas were supported by the ACEE task force, and in September 1975, Edgar Cortright, the Langley Director, initiated the Laminar-Flow-Control Working Group. Cortright announced that Langley had accepted the responsibility of implementing a research and technology program focused on the "development and demonstration of economically feasible, reliable, and maintainable laminar flow control."[16] One of the primary new focuses was a change from military to commercial applications.

There seemed to be as many staunch proponents of laminar flow as there were detractors. The optimists believed that a laminar flow wing could be developed using existing manufacturing techniques and known materials and implemented in a reasonable timeframe: by the 1990s. The laminar flow pessimists argued that even if all these achievements were possible (and many believed they were not), the costs and efforts required to keep the airfoil surfaces smooth, clean, and in flight-ready condition would make the entire system prohibitive. The airline industry summarized its concerns in four main areas: manufacturability, operational sensitivity, maintainability, and methodology.[17] Hans Mark, the Director of Ames Research Center, was one detractor. He said that the laminar flow program under ACEE should be "given low priority due to the low probability of success, and because benefits are not likely to be realized for many years, if ever."[18]

The laminar flow group within ACEE had a difficult mission in front of it: to provide data to support or refute assumptions by both the optimistic and pessimistic camps so that industry could make "objective decisions on the feasibility of laminar flow control for application to commercial

---

16. Edgar Cortright, "Establishment of Laminar Flow Control Working Group," Sept. 12, 1975, as found in Braslow, *A History of Suction-Type Laminar Flow Control,* p. 61.
17. "ACEE Program Overview," NASA RP-79-3246(1), July 31, 1979, Box 239, Division 8000, NASA Glenn archives.
18. Mark to Lovelace, June 4, 1975, Box 181, Division 8000, NASA Glenn archives.

transports of the 1990s time period."[19] Despite the uncertainties, laminar flow was included in ACEE for two main reasons: first, it offered the promise of dramatic fuel-efficiency improvement, and second, the work in composites might directly contribute to developing materials more operationally and economically suited for achieving laminar flow control.

The program, "involved a major change in Agency philosophy regarding aeronautical research," according to Albert Braslow. It included an extension of the traditional NACA role in research to include a "*demonstration* of technological maturity in order to stimulate the application of technology by industry."[20] This was also a risky proposition, made even more so during the political environment of the Reagan years. Project managers accepted the high level of risk in taking on this program because it was such a revolutionary idea with such great potential. Because NASA had to produce flight research results in several areas, it decided that a phased approach—by breaking down the problems into smaller units—would offer the best chances of success. Phase one involved developing methods for analyzing boundary layers with new computer codes. Also included were studies of surface materials and how to best maintain them. Phase two would move to basic fabrication of test pieces and subject them to wind tunnel testing. This would include subsystems such as pumps for suctioning. Phase three included actual flight-testing, with laminar flow control over a wing or a tail. Braslow was extremely enthusiastic about the potential for the program but was also aware of the risk. He said, "Everybody agrees that you have a hell of a payoff, but the question is, 'Can you do it on a day-to-day basis?'"[21]

As phases one and two progressed, several key problems were overcome. Insect contamination was thought by many to be a critical issue in preventing program success. Although the insect remains on the wings were small, they were nonetheless large enough to disrupt laminar flow. That an insect represented the margin of success or failure suggests how difficult the project was. Engineers tested washing systems and nonstick surface materials and concluded that it was best to keep the wings wet

---

19. Braslow and Muraca, "A Perspective of Laminar-Flow Control," *AIAA Conference on Air Transportation, Aug. 21–24, 1978*, p. 14.

20. Emphasis in original. Braslow, *A History of Suction-Type Laminar Flow Control,* p. 15.

21. Braslow, quoted in Wetmore, "Langley Presses Fuel Efficiency Programs," *Aviation Week & Space Technology,* Nov. 10, 1975, p. 68.

so the insects they encountered wouldn't stick.[22] The potential impact of engine-generated noise waves disrupting laminar flow on wings was another area of concern, and a NASA contract with Boeing investigated the laminar flow acoustic environment on a 757. Engine noise, it was found, did not cause the laminar flow to become turbulent. Research went beyond suction laminar flow control. Natural laminar flow investigations were carried out on F-111 and F-14 jets at Dryden Flight Research Center.

With success in these first two phases building confidence, phase three began by selecting a vehicle for flight-testing. The airlines wanted an aircraft similar in size to their commercial transports, while NASA pushed for a smaller plane to reduce costs. A compromise was eventually made using a larger plane but restricting experiments to the leading edge of a laminar flow wing, the most technically difficult area to overcome. The leading edges had to be smoother than other areas and had to withstand rain, insects, corrosion, icing, etc. Langley eventually used a JetStar plane, similar in size to a DC-9. NASA contracted with three industry leaders—Douglas, Lockheed, and Boeing—with NASA assuming 90 percent of the cost.

The Lockheed studies used a composite (graphite epoxy) wing covered by a very thin titanium sheet. The ducting was achieved through slots, and compressors induced the suction. However, it forced the wing to maintain the entire weight of the system, which became problematic. Douglas engineers used a different approach, opting for perforated holes instead of slots for the ducting, and explored using a glass fiber material for the suctioning. Boeing came to the laminar flow studies later than the other two companies, preferring to focus all its early attention on near-term fuel efficiency endeavors, as opposed to the uncertain future of laminar flow control.[23]

After 4 years of flight tests (1983 to 1987), all results were extremely positive.[24] Laminar flow control had been achieved for this leading edge area of the wing in a variety of test conditions, including cold, heat, rain,

---

22. David F. Fisher and John B. Peterson, Jr., "Flight Experience on the Need and Use of Inflight Leading Edge Washing for a Laminar Flow Airfoil," *AIAA Aircraft Systems and Technology Conference* (AIAA Paper 78-1512, Aug. 21–23, 1978).
23. "Laminar Flow Research Enters Tunnel, Flight Test," *Aviation Week & Space Technology,* Sept. 25, 1978, p. 49.
24. Dal V. Maddalon and Braslow, *Simulated-Airline-Service Flight Tests of Laminar-Flow Control with Perforated-Surface Suction System* (Washington, DC: NASA TP-2966, Mar. 1990).

Laminar flow test aircraft in flight (November 15, 1984). (NASA Langley Research Center [NASA LaRC].)

freezing rain, ice, moderate turbulence, and insects. Pilots had no difficulty adjusting to the new system. The titanium surface did not corrode over time. Enthusiasm soared higher after a series of test flights with the C-140 JetStar at Ames-Dryden Flight Research Facility, which simulated a commercial airline service operating in a variety of weather conditions and achieved 22-percent fuel efficiency at cruise speed. Roy Lange, the Laminar Flow Control program manager at Lockheed-Georgia, was pleased with the initial results, though more work still awaited completion. "The only question we have now," he said in 1985, "is whether the systems can handle a day-by-day flight schedule. . . . I think we could get there for a 1995 aircraft."[25] In addition, Langley engineers also investigated hybrid laminar flow control, a combination of the suction and natural laminar flow techniques. Boeing began research on a 757.[26] Braslow recalled that "results were very encouraging. . . . All necessary systems required for practical [hybrid laminar flow control] were successfully installed into a commercial transport wing."[27] Calculated benefits for a 300-person transport predicted a 15-percent savings in fuel.

---

25. Roy Lange, quoted in Keith F. Mordoff, "NASA C-140 with Laminar Flow Wing Simulating Airline Service Flights," *Aviation Week & Space Technology,* Apr. 15, 1985, p. 58.
26. Chambers, *Innovation in Flight.*
27. Braslow, *A History of Suction-Type Laminar Flow Control,* p. 32.

Despite the successful outcomes, laminar flow control is not currently used in any commercial transport. While the concept was proved in theory and flight-tested, it was never put into service nor put through the rigors of a day-to-day operational environment. It fell victim to the drop in fuel prices in the late 1980s, as there was no economic incentive for pushing through the remaining technological obstacles and actually incorporating laminar flow control into a commercial airlines' service.

There has been some continued laminar flow research that has yielded positive results since the end of ACEE, including the NASA–Boeing–Air Force B-757 Hybrid Laminar Flow Control (HLFC) flight experiments. As one Langley press release noted in August 1990, the "aerodynamic efficiency of future aircraft may improve sharply due to better-than-expected findings from a joint-government-industry flight test program." Laminar flow was achieved over 65 percent of the modified 757 wing, and engineers speculated that if the entire span of both of the wings were modified, the airplane drag would decrease by 10 percent. This would save roughly $100 million annually for the U.S. airline industry.[28] Despite the progress, the technology was not perfected. In 2004, aeronautical engineers William S. Saric and Helen L. Reed presented a paper on the remaining challenges in achieving practical laminar flow. They concluded that "crossflow instability" remained the most significant challenge.[29]

Richard Wagner, the head of the program, lamented the fact that the laminar technology is still unused. He said, "I really was disappointed that we didn't see, or haven't seen an application of . . . laminar flow control because . . . the stuff was ready. I guess it's just going to take some time to where the fuel price makes it so attractive that they can't turn their back on it."[30] Despite its lack of industry acceptance, the ACEE program made major advances in understanding the potential of laminar flow. As James Hansen argued, "all of the promising research indicated that its time might yet come."[31]

---

28. H. Keith Henry, "Flight Tests Prove Concept for Jetliner Fuel Economy," Aug. 23, 1990, as found at *http://www.scienceblog.com/community/older/archives/D/archnas1334.html,* accessed June 1, 2009.
29. William S. Saric and Helen L. Reed, "Toward Practical Laminar Flow Control—Remaining Challenges," AIAA Fluid Dynamics Conference, June 28–July 1, 2004, Portland, OR, as found at, *http://flight.tamu.edu/pubs/papers/aiaa-2004-2311.pdf,* accessed June 1, 2009.
30. Interview with Wagner by Bowles, June 30, 2008.
31. Hansen, *The Bird Is on the Wing,* pp. 204–205.

The Wave of the Future—Advanced Turboprop Project

Like many of the ACEE projects, the turboprop's history started before ACEE was born.[32] The project began in the early 1970s with the collaboration of two engineers: Daniel Mikkelson, of NASA Lewis, and Carl Rohrbach of Hamilton Standard, the Nation's last propeller manufacturer. Mikkelson, then a young aeronautical research engineer, went back to the old NACA wind tunnel reports, where he found a "glimmer of hope" that propellers could be redesigned to make propeller-powered aircraft fly faster and higher than did those of the mid- to late 1950s.[33] Mikkelson and Rohrbach came up with the concept of sweeping the propeller blades to reduce noise and increase efficiency. At Lewis, Mikkelson sparked the interest of a small cadre of engineers, who solved key technological problems essential for the creation of the turboprop, while at the same time attracting support for the project. The engineers also became political advocates, using their technical gains and increasing social acceptance to fight for continued funding. This involved winning Government, industry, and public acceptance for the new propeller technology. While initially the project involved only Hamilton Standard, the aircraft engine manufacturers—Pratt & Whitney, Allison, and General Electric—and the giants of the airframe industry—Boeing, Lockheed, and McDonnell-Douglas—jumped on the bandwagon as the turboprop appeared to become more and more technically and socially feasible. The turboprop project became a large, well-funded, "heterogeneous collection of human and material resources" that contemporary historians refer to as "big science."[34]

At its height, it involved over 40 industrial contracts, 15 university grants, and contracts with all 4 NASA Research Centers: Lewis, Langley, Dryden, and Ames. The project nonetheless remained controversial through its life, because of technical and social challenges. Technically, studies by Boeing, McDonnell-Douglas, and Lockheed pointed to four

---

32. Much of this section appeared in Bowles and Dawson, "The Advanced Turboprop Project: Radical Innovation in a Conservative Environment," *From Engineering Science to Big Science: The NACA and NASA Collier Trophy Research Project Winners,* Pamela E. Mack, ed., (Washington, DC: NASA SP-4219, 1998), pp. 321–343.
33. Interview with Dan Mikkelson by Dawson and Bowles, Sept. 6, 1995.
34. James H. Capshew and Karen A. Rader, "Big Science: Price to the Present," *Osiris,* 2nd ser., vol. 7 (1992), pp. 3–25.

John Klineberg, right, and Andy Stofan in 1983 with an Advanced Turboprop model. (NASA Glenn Research Center [NASA GRC].)

areas of concern: propeller efficiency at cruise speeds, both internal and external noise problems, installation aerodynamics, and maintenance costs.[35] Socially, the turboprop also presented daunting problems. Because of the "perception of turboprops as an old-fashioned, troublesome device with no passenger appeal," the consensus was that, "the airlines and the manufacturers have little motivation to work on this engine type."[36]

The project had four technical stages: "concept development" from 1976 to 1978, "enabling technology" (1978 to 1980), "large scale integration" (1981 to 1987), and finally "flight research" in 1987.[37] During each of

35. Roy D. Hagar and Deborah Vrabel, *Advanced Turboprop Project* (Washington, DC: NASA SP-495, 1988), p. 5.
36. "Aircraft Fuel Conservation Technology Task Force Report," Office of Aeronautics and Space Technology, Sept. 10, 1975, p. 44.
37. Hagar and Vrabel, *Advanced Turboprop Project,* pp. 6–10.

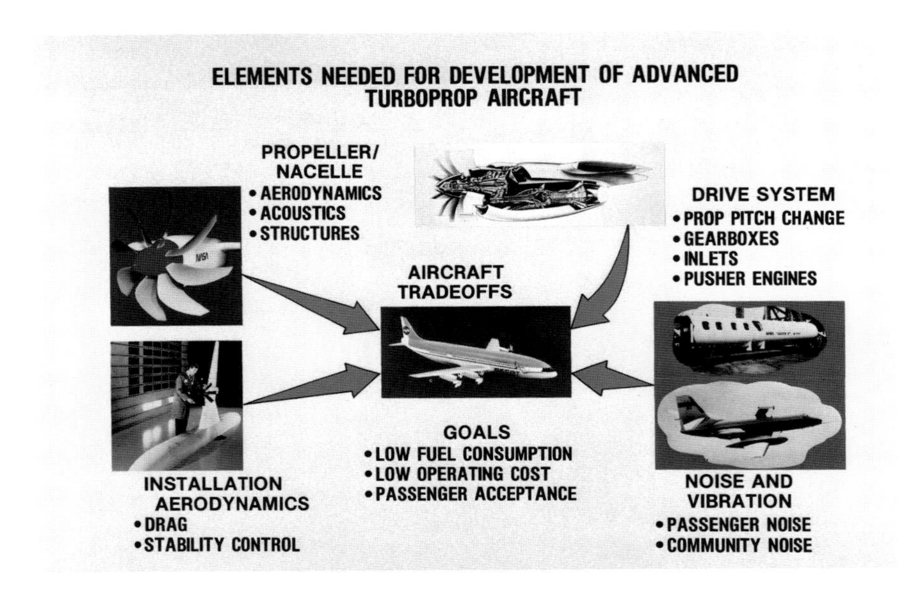

Advanced Turboprop elements, including the propeller, nacelle, aerodynamics, and noise control. (NASA Glenn Research Center [NASA GRC].)

these stages, NASA's engineers confronted and solved specific technical problems that were necessary if the Advanced Turboprop project were to meet the defined Government objectives concerning safety, efficiency, and environmental protection. Industry resistance and NASA Headquarters' sensitivity to public opposition were among the key reasons that of the six projects within the ACEE program, only the Advanced Turboprop failed to receive funding in 1976. John Klineberg, Director of Lewis Research Center, recalled that it was delayed "because it was considered too high risk and too revolutionary to be accepted by the airlines."[38] Everyone, it seemed, associated the advanced turboprop technology with the possibility of inciting an aeronautical "revolution," a paradigm shift, or, as a *Forbes* magazine headlined it in 1981, "The Next Step." As surely as "jets drove propellers from the skies," the new "radical designs" could bring a new propeller age to the world.[39] Donald Nored proclaimed that they were the "wave of the future."[40]

---

38. Klineberg, quoted in "How the ATP Project Originated," *Lewis News,* July 22, 1988.
39. Howard Banks, "The Next Step," *Forbes,* May 7, 1984, p. 31.
40. Nored to G. Keith Sievers, Jan. 9, 1981, Box 260, Division 8000, NASA Glenn archives.

Unfortunately, the airline industry was reluctant to return to the propeller. According to Nored, executives in the industry were "very conservative, and they had to be." They were "against propellers" because they had "completely switched over to jets." Because of their commitment to the turbojet, they raised numerous objections to a new propeller, including noise, maintenance, and the fear that the "blades would come apart." Nored recalled each problem had to be "taken up one at a time and dealt with."[41] The revolutionary propeller-driven vision of the future frightened the aircraft industry with its large investment in turbofan technology. Aircraft structures and engines are improved in slow, conservative, incremental steps. To change the propulsion system of the Nation's entire commercial fleet represented an investment of tremendous proportions. Even if the Government put several hundred million dollars into developing an advanced turboprop, the airframe and aircraft engine industries would still need to invest several billion dollars more to commercialize it. Revolutionary change did not come easily to an established industry so vital to the Nation's economy.

While fuel savings between 20 to 30 percent were one reason to take this risk, another important political factor favored its development. The Soviet Union had a "turboprop which could fly from Moscow to Havana."[42] The continuing Cold War prompted the United States to view any Soviet technical breakthrough as a potential threat to American security. During the energy crisis, the knowledge that Soviet turboprop transports had already achieved high propeller fuel efficiency at speeds approaching those of jet-powered planes seemed grave indeed and gave impetus to the NASA program. During the Government hearings, NASA representatives displayed several photos of Russian turboprop planes to win congressional backing for the project.[43] The Cold War helped to define the turboprop debate. No extensive speculation on the implications of Russian air

---

41. Interview with Nored by Dawson and Bowles, Aug. 15, 1995.

42. Interview with Mikkelson by Dawson and Bowles, Sept. 6, 1995.

43. "Aircraft Fuel Conservation Technology Task Force Report," Office of Aeronautics and Space Technology, Sept. 10, 1975, p. 48. These Soviet long-range turboprops included the Tupolev TU-95 "Bear" (which weighed 340,000 pounds, had a maximum range of 7,800 miles and a propeller diameter of 18.4 feet, and operated at a 0.75 Mach cruise speed) and the Antonov AN-22 "Cock" (which weighed 550,000 pounds, had a maximum range of 6,800 miles and a propeller diameter of 20.3 feet, and operated at a 0.69 Mach cruise speed).

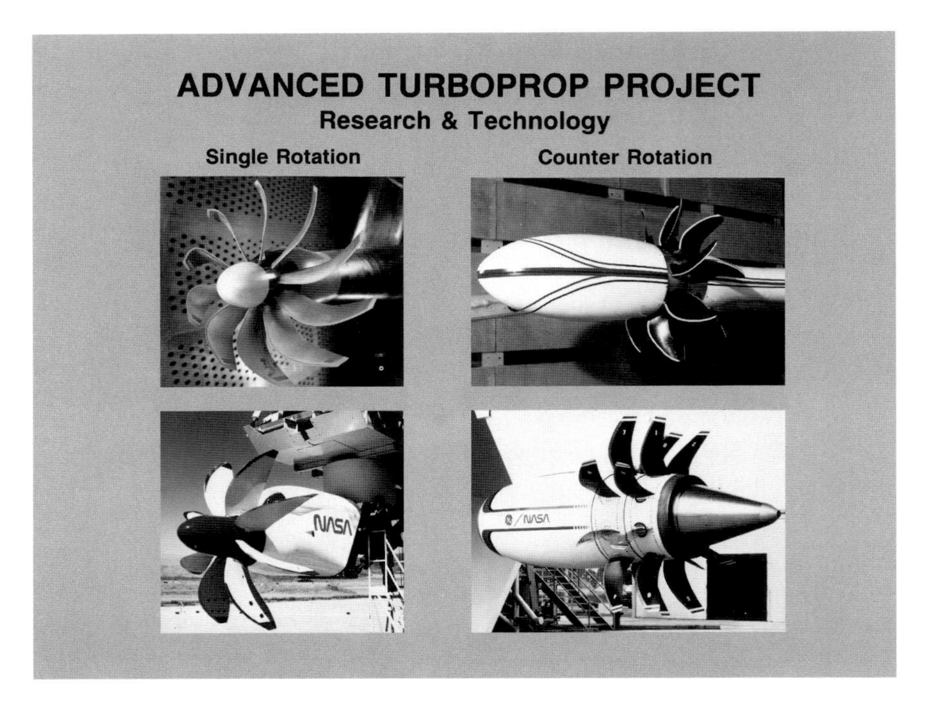

Single- and counter-rotation turboprops. (NASA Glenn Research Center [NASA GRC].)

superiority for American national security seemed necessary. The Soviet Union could not be allowed to maintain technical superiority in an area as vital as aircraft fuel efficiency.

The first step in developing a turboprop was to create a small-scale model. Technically, the entire future of the Advanced Turboprop project initially depended on proving whether a model propfan could achieve the predicted fuel-efficiency rates. If this model yielded success, then project advocates would be able to lobby for increased funding for a large research and development program. Thus, even during its earliest phase, the technical and social aspects of the project worked in tandem. Lewis project managers awarded a small group of researchers at Lewis and Hamilton Standard a contract for the development of a 2-foot-diameter model propfan, called the SR-1, or single-rotating propfan. Single-rotating meant that the propfan had only one row of blades, as opposed to a counter-rotating design with two rows of blades, each moving in opposite directions. This model achieved high efficiency rates and provided technical data that the small group of engineers could use as ammunition in the fight to continue the program.

This success led to the formal establishment of the program in 1978 and the enabling technology phase. Technically, this phase dealt with four critical problems: modification of propeller aerodynamics, cabin and community noise, installation aerodynamics, and drive systems. Propeller aerodynamic work included extensive investigations of blade sweep, twist, and thickness. In the late 1970s, for the first time, engineers used computers to analyze the design of a propeller. The advantage of propellers in saving fuel had to be balanced against a potential increase in noise pollution.[44] New computer-generated design codes not only contributed to improved propeller efficiency, but also to solving many of the problems associated with noise. The final two technical problems of the enabling phase dealt with installation aerodynamics and the drive system. Numerous installation arrangements were possible for mounting the turboprop on the wing. Should the propeller operate by "pushing" or "pulling" the aircraft? How should the propeller, nacelle, and the wing be most effectively integrated to reduce drag and increase fuel efficiency? Wind tunnel tests reduced drag significantly by determining the most advantageous wing placement for the propeller. Engineers also examined various drive train problems, including the gearboxes.

After 2 years of work, the turboprop idea began to attract greater commercial and military interest and support. The Navy's assistant commander for research and technology planned to incorporate it as a "viable candidate" for future long-range and long-endurance missions.[45]

A Lockheed-California vice president lent his support to the project, saying it would result in performance improvement for military application and "provide important means for future energy conservation in air transportation."[46] In 1978, the vice president for engineering at United Airlines reported that, after the company's management review on the ACEE turboprop project, it was "impressed with the progress made to date and the promise for the future."[47] One year later, United Airlines

---

44. "Aircraft Fuel Conservation Technology Task Force Report," Office of Aeronautics and Space Technology, Sept. 10, 1975, pp. 18, 46.

45. C.R. Copper to Kramer, Mar. 5, 1979, Box 179, Division 8000, NASA Glenn archives.

46. Lloyd E. Frisbee to Howard Cannon, Chairman of the Commerce, Science and Transportation Senate Committee, Feb. 23, 1979, Box 179, Division 8000, NASA Glenn archives.

47. Robert C. Collins to Robert A. Frosch, Dec. 1, 1978, Box 179, Division 8000, NASA Glenn archives.

president Percy A. Wood reiterated support for the program. Wood was "impressed" with the program and believed it was of "utmost importance" for the Nation and would have a "major impact" on the future of air transportation.[48]

With the small-scale model testing complete and growing industry support, the project moved into its most labor- and cost-intensive phase—large-scale integration. The project still had serious uncertainties and problems associated with transferring the designs from a small-scale model to a large-scale propfan. The Large-Scale Advanced Propfan (LAP) project initiated in 1980 would answer these scalability questions and provide a database for the development and production of full-size turbofans. As a first step, NASA had to establish the structural integrity of the advanced turboprop.[49] Project managers initially believed that in the development hierarchy, performance came first, then noise, and finally structure. As the project advanced, it became clear that structural integrity was the key technical problem.[50] Without the correct blade structure, the predicted fuel savings could never be achieved. NASA awarded Hamilton Standard the contract for the structural blade studies that were so crucial to the success of the program. In 1981, Hamilton Standard began to design a large-scale, single-rotating propfan. Five years later, construction was completed on a 9-foot-diameter design very close to the size of a commercial model, which was so large that no wind tunnel in the United States could accommodate it. The turboprop managers decided to risk the possibility that the European aviation community might benefit from the technology NASA had so arduously perfected. They shipped the SR-7L to a wind tunnel in Modane, France, for testing. In early 1986, researchers subjected the model to speeds up to Mach 0.8 with simulated altitudes of 12,000 feet. The results confirmed the data obtained from the small-model propeller designs. The large-scale model was a success.

Another key concern was unrelated to technological capability. This was a social question concerning passengers: How receptive would they be to propeller-driven aircraft? Laminar flow control and supercritical

48. Percy A. Wood to Frosch, Sept. 14, 1979, Box 179, Division 8000, NASA Glenn archives.
49. "Large-Scale Advanced Prop-Fan Program (LAP)," technical proposal by Lewis Research Center, Jan. 11, 1982, NASA, Nored papers, Box 229, NASA Glenn archives.
50. Interview with Nored by Dawson and Bowles, Aug. 15, 1995.

airfoils could be integrated into an airframe design without the public realizing, for the most part, the technology was even there. Turboprops were different because the propeller was one of the more visible parts of the airplane and was regarded as being from the "old days" when noisy "puddle-jumpers" were flown at low altitudes in turbulence. If the public would not fly in a turboprop plane, all the efficiency savings would be lost flying empty planes across the country.

In response to this concern, NASA and United Airlines initiated an in-flight questionnaire to determine customer reaction to propellers. Both NASA and the industry were aware of the disastrous consequences for the future of the program if this study found the public opposed the return of propeller planes. As a result, the questionnaire deemphasized the propeller as old technology and emphasized the turboprop as the continuation and advancement of flight technology. The first page of the survey consisted of a letter from United Airlines' vice president of marketing to the passenger asking for cooperation in a "joint industry-government study concerning the application of new technology to future aircraft."[51] This opening letter did not mention the new turboprops. The turboprop, inconspicuously renamed the "prop-fan" to give it a more positive connotation, did not make its well-disguised appearance until page 4 of the survey, where the passenger was finally told that "'prop-fan' planes could fly as high, as safely, and almost as fast and smooth as jet aircraft." This was a conscious rhetorical shift from the term "propeller" to "prop-fan" to disassociate it in people's minds from the old piston engine technology of the pre–jet-propulsion era. Brian Rowe, a General Electric engineer involved in advanced propeller projects, explained this new labeling strategy. He said, "They're not propellers. They're fans. People felt that modern was fans, and old technology was propellers. So now we've got this modern propeller which we want to call a fan."[52] The questionnaire explained to the passenger that not only did the "'prop-fans'. . . look more like fan blades than propellers," they would also use 20 to 30 percent less fuel than jet aircraft did.

The questionnaire then displayed three sketches of planes—two were propeller driven, and the third had a turbofan. The passenger had to choose

---

51. "United Airlines Passenger Survey," Box 224, Division 8000, NASA Glenn archives.

52. Quoted by Hamilton, "Firms Give Propellers a New Spin: GE leads high-stakes competition for aircraft engineers with its 'fan,'" *Washington Post,* Feb. 8, 1987.

SR-7L Advanced Turboprop on gulfstream jet in 1987. (NASA Glenn Research Center.)

which one he or she would "prefer to travel in." Despite all the planes being portrayed in flight, the sketches depicted the propellers as simple circles (no blades present), while the individual blades of the turbofan were visible. These were all subtle and effective hints to the passenger that the "prop-fan" was nothing new and that they were already flying in planes powered by engines with fan blades.

Not surprisingly, the survey yielded favorable results for the turboprop. Of 4,069 passengers surveyed, 50 percent said they "would fly prop-fan," 38 percent had "no preference," and only 12 percent preferred a jet.[53] If the airlines could avoid fare increases because of the implementation of the turboprop, 87 percent of the respondents stated they would prefer to fly in the new turboprop. Relieved and buoyed by the results, NASA engineers liked to point out that most of the passengers did not even know what was currently on the wing of their aircraft.[54] According to Mikkelson, all the passengers wanted to know was "how much were the drinks, and how much was the ticket."[55] Equally relieved was Robert Collins, vice president of engineering for United Airlines, who concluded that this "carefully constructed passenger survey . . . indicated that a prop-fan with equivalent passenger comfort levels would not be negatively viewed, especially if it were recognized for its efficiency in reducing fuel consumption and holding fares down."[56]

Success spawns imitators. While NASA continued to work with Allison, Pratt & Whitney, and Hamilton Standard to develop its advanced turboprop, General Electric (Pratt & Whitney's main competitor) was quietly developing an alternative propeller system—the unducted fan (UDF). In NASA's design, the propeller rotated in one direction. This was called a single rotation tractor system and included a relatively complicated gearbox. Since one of the criticisms of the turboprop planes of the 1950s (the Electra, for example) was that their gearboxes required heavy maintenance, General Electric took a different approach to propeller design. Beginning in 1982, its engineers spent 5 years developing a gearless, counter-rotating pusher system. They mounted two propellers (or fans) on the rear of the plane that literally pushed it in flight, as opposed to the "pulling" of conventional propellers. In 1983, the aircraft engine division of General Electric released the unducted fan design to NASA, shortly before flight tests were scheduled.

This took NASA completely by surprise. Suddenly, there were two turboprop projects competing for the same funds. Nored recalled: "They

53. "Prop-Fan survey results," Dec. 1978, Box 231, Division 8000, NASA Glenn archives.
54. Interview with Sievers by Dawson and Bowles, Aug. 17, 1995.
55. Interview with Mikkelson by Dawson and Bowles, Sept. 6, 1995.
56. Robert C. Collins, statement submitted to the Subcommittee on Transportation, Aviation, and Materials, House of Representatives Committee on Science and Technology, Feb. 26, 1981.

The Dryden C-140 JetStar during testing of advanced propfan designs. Dryden conducted flight research in 1981–1982. The Lewis Research Center directed the technology's development under the Advanced Turboprop program. Langley oversaw work on acoustics and noise reduction. The effort was intended to develop a high-speed, fuel-efficient turboprop system. (January 1, 1981.) (NASA Dryden Flight Research Center [NASA DFRC].)

wanted us to drop everything and give them all our money, and we couldn't do that."[57] NASA Headquarters endorsed the "novel" unducted fan proposal and told Lewis to cooperate with General Electric on the unducted fan development and testing. Despite NASA's initial reluctance to support two projects, the unducted fan proved highly successful. In 1985, ground tests demonstrated a fuel-conservation rate of 20 percent. Development of the unducted fan leapt ahead of NASA's original geared design. One year later, on August 20, 1986, General Electric installed its unducted fan on the right wing of a Boeing 727. Thus, much to NASA engineers' dismay, the first flight of an advanced turboprop system demonstrated the technical feasibility of the unducted fan system—a proprietary engine belonging to entirely General Electric, not a product of the joint NASA-industry team. Nevertheless, the competition between the two systems, and the willingness of private industry to invest development funds, helped build even greater momentum for acceptance of the turboprop concept.

57. Interview with Nored by Dawson and Bowles, Aug. 15, 1995.

NASA engineers continued to perfect their single-rotating turboprop system through preliminary stationary flight-testing.[58] The first step was to take the Hamilton Standard SR-7A propfan and combine it with the Allison turboshaft engine and gearbox housed within a special tilt nacelle. NASA engineers conducted a static or stationary test at Rohr's Brown Field in Chula Vista, CA, mounting the nacelle, gearbox, engine, and propeller on a small tower. The stationary test met all performance objectives after 50 hours of testing in May and June 1986, a success that cleared the way for an actual flight test of the turboprop system. In July 1986, engineers dismantled the static assembly and shipped the parts to Savannah, GA, for reassembly on a modified Gulfstream II with an eight-blade, single-rotation turboprop on its left wing.[59] The radical dreams of the NASA engineers for fuel-efficient propellers were finally close to becoming reality. The plane contained over 600 sensors to monitor everything from acoustics to vibration. Flight-testing—the final stage of advanced turboprop development—took place in 1987, when a modified Gulfstream II took flight in the Georgia skies. These flight tests proved that the predictions NASA made in the early 1970s of a 20- to 30-percent fuel savings were indeed correct.

On the heels of the successful tests of both the General Electric and the NASA-industry team designs came not only increasing support for propeller systems themselves, but also high visibility from media reports predicting the next propulsion revolution. The *New York Times* predicted the "Return of the propellers" while the *Washington Times* proclaimed, "Turboprops are back!"[60] Further testing indicated that this propulsion technology was ready for commercial development. As late as 1989, the U.S. aviation industry was "considering the development of several new engines and aircraft that may incorporate advanced turboprop propulsion systems."[61] But the economic realities of 1987 were far different from those predicted in the early 1970s. Though all the problems standing in

---

58. Hagar and Vrabel, *Advanced Turboprop Project,* pp. 49–74. This stage was called the Propfan Test Assessment (PTA) project.
59. Mary Sandy and Linda S. Ellis, "NASA Final Propfan Program Flight Tests Conducted," *NASA News,* May 1, 1989.
60. Andrew Pollack, "The Return of Propellers," *New York Times,* Oct. 10, 1985, p. D2. Hugh Vickery, "Turboprops are back!" *Washington Times,* Nov. 1, 1984, p. 5B.
61. Sandy and Ellis, "NASA Final Propfan Program Flight Tests Conducted," *NASA News,* May 1, 1989.

the way of commercialization were resolved, the advanced turboprop never reached production, a casualty of the one contingency that NASA engineers never anticipated—fuel prices decreased. Once the energy crisis passed, the need for the advanced turboprop vanished. Oil cost $3.39 per barrel in 1970. It was $37.42 per barrel in 1980. By 1988, it had dropped to $14.87 per barrel, and ACEE programs such as Laminar Flow Control and the Advanced Turboprop lost their relevance.[62] As the energy crisis subsided in the 1980s and fuel prices decreased, there was no longer a favorable ratio of cost to implement turboprop technology versus savings in fuel efficiency. As John R. Facey, Advanced Turboprop Program Manager at NASA Headquarters, wrote, "An all new aircraft with advanced avionics, structures, and aerodynamics along with high-speed turboprops would be much more expensive than current turbofan-powered aircraft, and fuel savings would not be enough to offset the higher initial cost."[63]

Yet managers of the Advanced Turboprop program, such as Keith Sievers, were convinced that the NASA-industry team had made a significant contribution to aviation that ought to receive recognition. Although NASA won several Collier trophies, which are regarded as the most prestigious award given annually for aerospace achievement for innovations related to the space program, it had produced no winners in aeronautics for almost 30 years. If the turboprop could win such an honor, it might justify the importance of this work. In hopes of winning the Collier Trophy, Sievers began mobilizing the aeronautical constituency that had participated in turboprop development. Although NASA Headquarters initially expressed some reluctance to press for the prize for a technology that was unlikely to be used, at least in the near future, the timing was perfect. There was little competition from NASA's space endeavors, since staff members in the space directorate were still in the midst of recovering from the Challenger explosion. As a result, in 1987 the National Aeronautic Association awarded NASA Lewis and the NASA-industry Advanced Turboprop team the Collier Trophy at ceremonies in Washington, DC, for developing a new fuel-efficient turboprop propulsion system.[64] The winning team

---

62. Historical inflation adjusted price data as found at *http://www.inflationdata.com/inflation/Inflation_Rate/Historical_Oil_Prices_Table.asp,* accessed Sept. 2, 2009.
63. John R. Facey, "Return of the Turboprops," *Aerospace American* (Oct. 1988), p. 15.
64. Citation for the Collier Trophy in Roy D. Hager and Deborah Vrabel, p. vi.

John Klineberg holding the Collier Trophy (May 13, 1988). (NASA Glenn Research Center [NASA GRC].)

included Hamilton Standard, General Electric, Lockheed, the Allison Gas Turbine Division of General Motors, Pratt & Whitney, McDonnell-Douglas, and Boeing—one of the larger and more diverse groups to be so honored in the history of the prize.

Some specific technologies that were designed for the turboprop project are in use today. These include noise reduction advances, gearboxes that use the turboprop design, and solutions to certain structural problems, such as how to keep the blades stable.[65] Today, the technology remains "on the shelf," or "archived," awaiting the time when fuel conservation again becomes a necessity. When interviewed in the mid-1990s, NASA engineers involved in the Advanced Turboprop Project remained confident that future economic conditions would make the turboprop attractive again. When fuel prices rise, the turboprop's designs will be "on the shelf," ready to provide tremendous fuel-efficient savings. But NASA engineers did not build their careers around technologies that were ultimately neglected. Donald Nored sentimentally reflected on the project, waved goodbye to the future of turboprops, and said, "We almost made it. Almost made it."[66]

---

65. Interview with Sievers by Dawson and Bowles, Aug. 17, 1995.
66. Interview with Nored by Bowles, Aug. 15, 1995.

EPILOGUE

FROM SHOCK TO TRANCE

"How quickly we forget our history," wrote newspaper editor Greg Knill in July 2008. Writing at a time of surging gas prices, with the price of oil reaching $147 per barrel, he reminded his readers of a time in the early 1970s when Middle East tensions drove up oil prices and fundamentally changed the way the Nation perceived the oil commodity. The United States Government called for energy independence, fuel economy was "all the rage," and the automobile industry reinvented itself with a switch to smaller cars. But, as Knill observed, there is an ebb and flow to everything, and the fuel crisis of the 1970s was replaced by perceptions of an oil glut in the 1980s. Fuel economy slowly disappeared in consumer purchasing decisions. Automobiles became larger and less efficient. And now, he wrote, the Nation faced a new crisis, which has emerged from the same global tensions and over the same finite world resource. Knill's response was: "The question I have is why the surprise?" and "how long will this current reawakening to the importance of fuel efficiency last?" He concluded, "Our history is not very encouraging."[1]

In July 2008, leaders in the American aviation community made a plea to President George W. Bush and Congress to call a special session to discuss the "full-blown and deepening energy crisis which is causing irreparable harm." Robert Crandall, the CEO of American Airlines, said that "our national confidence has been eroded," and that the rest of the world perceived that the United States "lacks the political will to address the energy crisis."[2] Although the problems facing the Nation in 2008 were eerily similar to those facing it in the early 1970s, other aviation experts

---

1. Greg Knill, "This is Not Our First Energy Crisis," *The News,* July 30, 2008.
2. Alexandra Marks, "Aviation Leaders Urge Congress to Act on Energy Policy—Now," *The Christian Science Monitor,* July 30, 2008.

realized that the problems was not just skyrocketing oil prices. The head of one airline industry analysis firm explained that the current crisis is tied to problems and decisions extending back 30 years or more. The airlines industry is cyclical, and during the good times, it is not profitable enough to prepare for future downturns or invest enough to fix its flaws. Just as in the early 1970s, the aviation industry and the Government began meeting to discuss the crisis and determine how best to plot the American response to it.

The frequency of these meetings increased by the day during 2008. In July, the American Association of Airport Executives held a summit, "The Energy Crisis and its Impact on Air Service," that convened experts from Congress, the Federal Aviation Administration, the airlines, and the Department of Transportation. The goal was to bring the aviation industry together with policy makers in Government "in an effort to frame the problem and work together to face these challenges."[3] A representative from the Air Transport Association said, "I think there is no greater crisis, not just for the airline industry, but across the board, than the energy crisis facing this country right now."[4] The Air Transport Association called upon Congress to assist in finding a bipartisan solution and establish reforms to help the struggling airline industry.[5]

None of this is new for those who remember the panic of the 1970s and especially for those who, in its wake, devoted their lives to developing fuel-efficient technologies for airplanes. Richard Wagner, one of the leaders of the NASA Langley Research Center's ACEE programs, said in June 2008: "I was amused a few weeks back when they announced that the airlines were slowing down to save fuel. Well, if they just slowed down to 0.8 Mach [we could take advantage of this technology]. It could have a natural laminar flow wing. It could have your turboprop propulsion. It's kind of amusing when you think about it. The steps that they're taking now, if they had taken those steps 10 years ago, they'd have very efficient aircraft flying around."[6]

---

3. *Defense and Aerospace Weekly,* June 23, 2008.
4. James May from the Air Transport Association, as found in Jamie Orchard, "Oil Crisis," *Global News Transcripts,* July 11, 2008.
5. "Air Transport Association Leads Coalition in Call for Bi-Partisan Near-Term Solutions to Energy Crisis," *Energy Weekly News,* June 23, 2008.
6. Wagner interview by Bowles, June 30, 2008.

The problem is developing a long-term energy plan that does not fluctuate with the changing price of oil and the changing demands of the market. When adjusting for inflation and using 2007 as a basis point, the price of a barrel of oil in 1970 was $18.77. By 1980, the price had risen to $97.68 per barrel. In these economic conditions, the ACEE program remained viable, and expensive fuel-saving technologies like Laminar Flow Control and the Advanced Turboprop were worth the investment. But by 1988, at the end of the ACEE project, the cost of a barrel of oil had fallen to $27.05. In that climate, there was no economic incentive to try to incorporate a revolutionary new airframe or propulsion system for commercial aviation. Prices continued to fall, and by 1998, a barrel of oil actually cost $3 less (in inflation-adjusted terms) than it did in 1970.[7] In summer 2008, the cost had risen to more than $140 per barrel.

If a graph plotted the price of oil and the Nation's interest in fuel efficiency together, the resulting curves would rise and fall at the same rate. The increased energy costs of the 1970s gave life to spreading new energy awareness, conserving fuel, lowering automobile speed limits, and establishing the Aircraft Energy Efficiency program. On our conceptual chart, we would see a peak in efficiency interest and oil price. During the 1980s and 1990s, with the perceptions of oil abundance, prices decreased, and the Nation entered into a collective amnesia about the importance of efficiency. NASA was seduced once again by the allure of "higher, faster, farther," and returned a High Speed Research (HSR) program at Langley.

This drive for faster aircraft has ebbed and flowed as nearly a mirror image of the desire for fuel efficiency. One example has been the longstanding goal to develop a Supersonic Transport (SST). It has been technologically feasible to develop a plane that travels faster than sound since the 1950s. Since that time, the Government has made three attempts to produce them for commercial use. The first was when the Kennedy Administration approved funding for a "national SST" program, but this was terminated in 1971, 2 years before the energy crisis. A second attempt, the Supersonic Cruise Aircraft Research (SCAR) program,

---

7. Historical inflation adjusted price data as found at *http://www.inflationdata.com/inflation/Inflation_Rate/Historical_Oil_Prices_Table.asp,* accessed Sept. 2, 2009.

NASA research engineer Dave Hahne inspects a tenth-scale model of a Supersonic Transport model in the 30- by 60-foot tunnel at Langley. The model is being used in support of NASA's High Speed Research program. (NASA Langley Research Center [NASA LaRC].)

was a smaller program that NASA hoped to have flight-ready by the 1980s. Funding for this terminated in 1981. Finally, 2 years after the ACEE program ended, the High Speed Research program commenced. This too was terminated, meeting its demise in 1998. Eric Conway has expertly told this story in his book *High-Speed Dreams*. "The long SST saga," he wrote, "reveals how national politics and business interest interact in the realm of high technology. All three American SST

Advanced Subsonic Technology test apparatus for combined bending and membrane test at Langley (November 7, 1997). (NASA Langley Research Center [NASA LaRC].)

programs were rooted in national and international politics. . . . All three collapsed when their political alliances disintegrated."[8]

Another collapsed aeronautics effort that lived and died alongside HSR was the Advanced Subsonic Technology (AST) program. AST was an intellectual offspring of ACEE and, along with High Speed Research,

8. Conway, *High-Speed Dreams,* pp. 1–2.

Bypass ratio 5 separate flow nozzle with chevron noise suppression trailing edge, photographed at Langley (October 20, 1999). (NASA Langley Research Center [NASA LaRC].)

was one of NASA's major aeronautics programs, with funding of $434 million for both.[9] AST commenced in 1993 and explored combustor emissions, fuel efficiency, composites technology, and noise reduction research through a Government-industry team that included NASA, GE Aircraft Engines, Pratt & Whitney, Allison Engines, and AlliedSignal Engines. Wesley Harris, NASA's Associate Administrator, told the House of Representatives in 1994 that NASA's objective in this program was to "provide US industry with a competitive edge to recapture market share, maintain a strongly positive balance of trade, and increase US jobs."[10] Though in theory this was an

9. Annual Meeting of the Transportation Research Board, *Transportation Demand Management and Ridesharing: [Papers Contained in This Volume Were Among Those Presented at the 76th TRB Annual Meeting in January 1997],* Transportation research record, 1598 (Washington, DC: National Academy Press, 1997), p. 18.

10. Wesley Harris, quoted in Philip K. Lawrence and David Weldon Thornton, *Deep Stall: The Turbulent Story of Boeing Commercial Airplane* (Aldershot, England: Ashgate, 2005), p. 95.

important program, its funding support was short-lived. AST was terminated in 2000, and many of the resources that had been allocated to it went to nonaeronautics programs, such as the International Space Station. According to one report, AST was terminated to "provide greater focus on public goods issues that threaten to constrain air system growth, such as aviation safety, airport delays, and aircraft emissions."[11]

AST was replaced in 2000 with the Ultra Efficient Engine Technology (UEET) program, which had "fewer resources and less industry involvement."[12] Its mission is to develop and then transfer to industry "revolutionary turbine engine propulsion technologies." The goals of this technology will be to address two of the more important propulsion issues—fuel efficiency and reduced emissions—which will lead to reducing ozone depletion and decreasing the role airplanes play in global warming.[13] The UEET program is managed by Glenn Research Center, with participation from other NASA Centers (Langley, Goddard, and Ames), engine companies (General Electric, Pratt & Whitney, Honeywell, Allison/Rolls-Royce, and Williams International), and airplane manufacturers (Boeing and Lockheed Martin). This team also collaborates with Government through relationships with the Department of Defense, the Department of Energy, the Environmental Protection Agency, and the Federal Aviation Administration.

As one might predict, with the ebbing interest in high-speed flight and increase in fuel prices, the incentive for fuel-efficient airplanes has returned. As fuel prices hit record highs with each passing day in mid-2008, the Nation once again scrambled to become energy conscious. Some of the old ACEE technology left to lie fallow in the 1980s is now being taken down from the shelf of deferred dreams. In June 2008, John E. Green, an engineer at the Aircraft Research Association in the United Kingdom, presented a paper at the AIAA Fluid Dynamics Conference titled "Laminar Flow Control—Back to the Future?" In it, he made a case

---

11. "Impact of the Termination of NASA's High Speed Research Program and the Redirection of NASA's Advanced Subsonic Technology Program," Report to the Congress, pp. 6–77, as found at *http://www.ostp.gov/pdf/hsr.pdf,* accessed June 1, 2009.
12. National Research Council (U.S.), *For Greener Skies: Reducing Environmental Impacts of Aviation,* (Washington, DC: National Academy Press, 2002), pp. 2, 31.
13. Robert J. Shaw, "UEET Overview," Tech Forum, Sept. 5–6, 2001, NASA Glenn archives.

Ultra Efficient Engine Technology (UEET) proof of concept compressor, two-stage compressor (December 4, 2003). (NASA Glenn Research Center [NASA GRC].)

for revisiting this old technology because full laminar flow control is possible based on more than 70 years of advances in aircraft engineering propulsion and materials. Green was aware that this was a risky topic for research because, as he said, "the level of interest in laminar flow has fluctuated with the price of oil, the price has never stayed high long enough to persuade any aircraft manufacturer to take the plunge." Citing tremendous advantages in fuel efficiency, and also reduced emissions that help improve the environment, Green implored his audience with a final plea: "Looking to the environmental and economic pressures that will confront aviation in the coming decades, we must conclude that it is now time to return in earnest to the challenge of building laminar flow control into our future transport aircraft."[14]

---

14. John E. Green, "Laminar Flow Control—Back to the Future?" *AIAA Fluid Dynamics Conference,* June 23–26, 2008, Seattle, WA, AIAA 2008-3738, as found at *http://adg.stanford.edu/aa241/supplement/Lam-Flow-Control-AIAA-2008-3738.pdf,* accessed June 1, 2009.

Glenn Research Center is also looking to the future by looking backward. Dennis Huff, the Deputy Chief of the Aeropropulsion Division at NASA's Glenn Research Center, began his career in 1985, just as the ACEE program was winding down. At the time, he heard a presentation by Bill Strack on the end of the ACEE Advanced Turbprop Project and wondered why the program was being canceled. Strack answered Huff's question by saying that itwas all about the price of fuel and predicted that if the fuel price ever tripled, some of these technologies would be taken off the shelf. Strack's prediction came true, as Glenn Research Center recently resumed work on an extension of the Advanced Turboprop Project, with a new counter-rotating open rotor that should be ready for commercial operation by 2015. Huff commented on the challenges of having changing national aeronautics priorities. Reflecting in summer 2008, he said, "It's been amazing over the last 2 years," when the priorities shifted from noise, to carbon dioxide emission, to fuel efficiency. Huff believed that NASA should work on all three and maintain a balanced approach, because he realized "there's no way we're going to change the market drivers and we can at least come up with the technology so people can make the choices they want to go with."[15] It remains to be seen whether Huff and his team will continue to have the support to complete their work.

In the wake of shifts in goals for civil aviation and an erosion of financial support, aeronautics as a whole continues to struggle for survival and funding at NASA. It never truly prospered after the "aeronautics wars" of the 1970s and 1980s, and the effects continue to be felt today. Many in the aviation industry believe the policies of the Reagan Administration have resulted in weakening the United States' position in the commercial transport market. One frequent visitor on Capitol Hill was Jan Roskam, an aircraft designer with Boeing and an aeronautics professor at the University of Kansas, who was often called by the House Committee on Science and Technology to provide testimony from 1974 to 1989. Roskam commented in 2002 about the effort to keep a distance between Government and the airlines industry. He wrote, "This very shortsighted decision has saved the taxpayers very little money and eventually will cost the U.S. its dominance in civil aeronautics."[16]

---

15. Dennis Huff, interview with Bowles, July 29, 2008.

16. Jan Roskam, *Roskam's Airplane War Stories: An Account of the Professional Life and Work of Dr. Jan Roskam, Airplane Designer and Teacher* (Lawrence, KS: DARcorporation, 2002), p. 134.

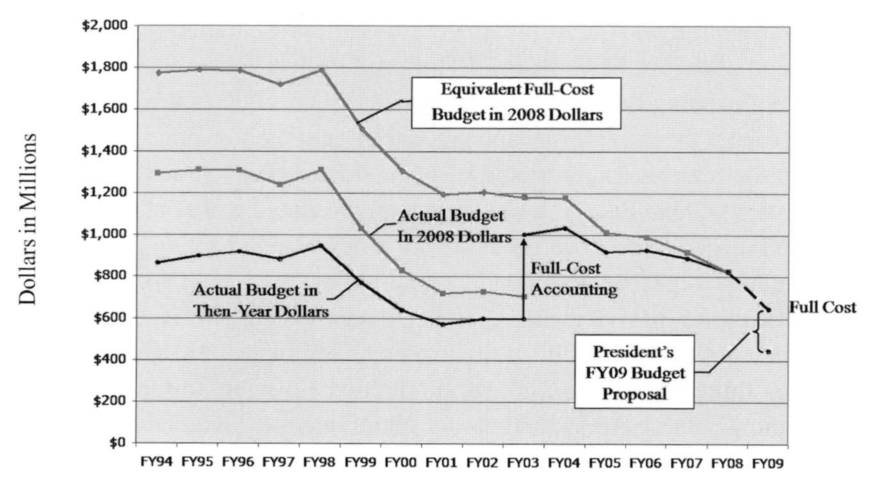

The above chart shows the decrease in NASA's aeronautics budget. Courtesy of Roy Harris, chief technical adviser to NASA's aeronautics support team. Not official NASA budget data.

Raymond Colladay recently argued, "Ever since NACA was morphed into NASA, the role of aeronautics in NASA has been kind of a stepchild."[17] Aeronautics, Colladay explained, has been threatened by so many cuts, and there is now no strong aeronautics advocate or leadership within the administration. "In the last 6 years or so," he said, "the aeronautics program has been continually cut to the point where it was on a glide-slope to go to zero." The cuts were difficult to fight because they were subtle and quiet, but not anymore. Colladay concluded, "It wasn't over any big ideological or political kind of battle which we faced in the eighties, it just was more from benign neglect."

Perhaps the era of neglect is over. Immediately after Barack Obama was elected President in November 2008, *60 Minutes* interviewed him, and he discussed the wild fluctuations in the price of oil. The question posed to him was: "When the price of oil was at $147 a barrel, there were a lot of spirited and profitable discussions that were held on energy independence.

---

17. Colladay, interview with Bowles, July 21, 2008.

Now you've got the price of oil under $60. . . . Does doing something about energy is it less important now. . . . " Before the interviewer had a chance to finish the question, Obama said, "It's more important. It may be a little harder politically, but it's more important." When asked why it is more important now when oil is so inexpensive, Obama explained that this was because this has been the pattern in recent American history. As oil prices go up, Obama explained, there is a political will to solve the problem. But he lamented that as soon as prices went back down, "suddenly we act like it's not important, and we start, you know, filling up our SUVs again." Obama called this phenomenon "from shock to trance" and said it was all a part of "our addiction." Obama concluded the interview with a vow that this pattern had to be broken: "Now is the time to break it."[18]

Breaking this pattern will be much more difficult, with NASA having a significantly weakened aeronautics capability thanks to years of low funding and support. Much of its DOD connections were cut during the administration of Daniel Goldin, which removed yet another funding opportunity. As Langley engineer Joseph Chambers observed, "Today's NASA aeronautics program is virtually invisible and without a large focused effort. Since the 1990s new embryonic starts are enthusiastically briefed to industry and DOD, advocacy gained within peer groups, and initiated with great fanfare, only to be canceled within a year or so."[19]

Although NASA, the Government, industry, and the public are all aware of the importance of fuel efficiency in this time of crisis, there is no guarantee that we have yet learned our lesson. As one reporter suggested in 2008, "If things settle down and seem less crisis-like in the future, are we going to lose interest? Will our attention spans be longer than they have been in the past? I don't know the answers to those questions."[20] Time will tell if we have become wiser and more able to enact policies and change public opinion that can endure longer than the latest oil price cycle. It will require a political will to impart a long-term vision to a country that

---

18. Andrew Revkin, "Obama on the 'Shock to Trance' Energy Pattern," Nov. 17, 2008, as found at, *http://dotearth.blogs.nytimes.com/2008/11/17/obama-on-shock-to-trance-energy-pattern,* accessed Sept. 2, 2009.
19. Chambers, correspondence to Bowles, Mar. 18, 2009.
20. Richard Mial, "If Energy Crisis Eases, Will We Stay Focused on Energy Issues?" *McClatchy-Tribune Business News,* May 29, 2008.

changes policy all too often to respond to immediate problems. What is needed is the will to break the trance and establish a lasting, structural, foundational plan to develop fuel-efficient aeronautics technologies. Does President Obama have this political will? At the time of this writing, it is still too early to tell. The only thing for certain is that the sooner we come to this understanding, the better. We cannot make more oil. It is a finite resource, the vast majority of it is not under our control, and it is the lifeblood of our economy. Only when we truly come to terms with this will we be ready to acknowledge the need and establish the long-term support and vision for a new "Apollo of Aeronautics" to help us escape from a predictable and lasting threat.

About the Author

Mark D. Bowles is associate professor of history at American Public University and founder of Belle History, a public history company (*www.bellehistory.com*). He has authored or coauthored eight books on the history of aeronautics, education, and medicine. His most recent, *Chains of Opportunity* (University of Akron Press, 2008), is a history of the emergence of the "polymer age" from the institutional perspective of one of the leading polymer universities in the world. His *Science in Flux* (NASA, 2006) won the American Institute of Aeronautics and Astronautics' 2005 History Manuscript Award, which is presented each year for the best historical manuscript dealing with the impact of aeronautics and astronautics on society. *The "Apollo" of Aeronautics* won the same award for 2010. He received his B.A. in psychology (1991) and M.A. in history (1993) from the University of Akron. He earned his Ph.D. in the history of technology (1999) from Case Western Reserve University and his MBA in technology management from the University of Phoenix (2005). He has been married to his wife, Nancy, for 19 years, and they are raising their 9-year-old daughter, Isabelle, and newborn twin girls, Emma and Sarah, in northeast Ohio. He can be reached at *mark@bellehistory.com*.

Index

C

D

E

GPO U.S. GOVERNMENT PRINTING OFFICE : 2010—359-890